3 R's of Nuclear Power: Reading, Recycling, and Reprocessing

...Making a Better Tomorrow for Little Joe

Dr. Jan Forsythe

authorHOUSE®

AuthorHouse™
1663 Liberty Drive, Suite 200
Bloomington, IN 47403
www.authorhouse.com
Phone: 1-800-839-8640

First published by AuthorHouse 11/2/2009

ISBN: 978-1-4389-6732-5 (sc)
ISBN: 978-1-4389-6731-8 (hc)

Library of Congress Control Number: 2009924874

Printed in the United States of America
Bloomington, Indiana

This book is printed on acid-free paper.

*This book is dedicated to the coming generations
who will inherit a complex and chaotic world.
Like my grandson Joe, they will either be blessed
or burdened by the decisions we make today.*

My legacy, hopefully, is a better tomorrow for Little Joe.

ABSTRACT

The author, Jan Forsythe, is a professor for the Florida Institute of Technology, teaching graduate courses in engineering management and systems engineering. Based on technical, scientific, and economic facts, her book is unique—a comprehensive story of nuclear power not heretofore told. Topics discussed include: "ABCs" of nuclear energy; experiences of the U. S. Navy and France in embracing nuclear power; history, current status, and future concepts; spent fuel recycling and final waste disposition; investments and operating costs; methodology and analysis, including tangible and intangible factors. Professor Forsythe analyzes the economics of nuclear power vs. coal, gas, and renewable energy sources to produce electricity. She reviews data, evaluates the information, and presents a positive vision for nuclear power in the United States.

AUTHOR'S NOTE

When I worked for Dr. Edward Teller, an advocate of nuclear energy, at The George Washington's Institute for Technology and Strategic Research, I became interested in nuclear power and the benefits that could be applied to peaceful endeavors, particularly as a source for generating electricity. I was captured by Dr. Teller's enthusiasm. It was contagious. When the Idaho National Laboratory offered me a position in the New Production Reactor Department, I jumped at the opportunity. I realized early on that radiation, security, and waste management were the three greatest concerns of the citizens of the United States. Therefore, I took my experience working with Dr. Teller and my colleagues at the Idaho National Laboratory, expanded by extensive research of nuclear information from newspapers, magazine articles, interviews, and books that have formed the public's fear of nuclear power, and placed it in a more rational mindset.

The research and analysis of this book are based on a working knowledge gleaned from participation in technical, planning, and strategy conferences and symposia attended over a period of 15 years. Sponsored by DoD, DOE, academia, and industry, the topical issues discussed included: national energy issues, nuclear science and technology, security and infiltration defenses of nuclear sites, industrial nuclear efficiency, proliferation of nuclear materials, transportation of nuclear materials and waste, nuclear energy and reactor efficiencies, nuclear cleanup and redevelopment, and geological disposal of nuclear waste.

No funding has been received for applied research, testing, or fabrication, as a result of or compensation for my personal participation and endeavors.

Jan I. Forsythe, D.Sc.
June 10, 2008

ACKNOWLEDGEMENTS OF APPRECIATION

To George R. Brier, D.Sc. Professor of Engineering Management and Systems Engineering, my skillful colleague in research, for the months we toiled to make this world a safer place to live. Dr. Brier, a retired Marine Corps general officer, was commanding general of the 4th Marine Aircraft Wing. Knowledgeable in nuclear technology and tactical nuclear weaponry from his military experience, he became increasingly interested in research of peaceful nuclear applications. He was a full-time professor at George Washington University for nearly 30 years. Dr. Brier, CEO of Learning Curve Coalition, Inc. is presently managing and teaching off-site engineering management programs for the Florida Institute of Technology.

My friend and colleague, Robert C. Howard, Principal Engineer, Idaho National Engineering Laboratory provides mechanical engineering management and technical support for operation, maintenance, and modifications to the advanced test reactor (ATR), the world's largest test reactor. An expert in reactor subsystem design, procurement, installation, and testing, he also was involved in the development and production of coated fuel particles. Mr. Howard not only reviewed my manuscript, but had colleagues at INL check it for technical accuracy.

Elwood P. Stroupe, Esquire, lawyer and expert in nuclear engineering, as Deputy Program Manager of Transportation of Yucca Mountain, assisted DOE in developing national nuclear transportation strategies, and is a technical expert on repository thermal issues. He was manager of the National Spent Nuclear fuel program in safety and reliability. A proponent for the potential beneficial uses of spent nuclear fuel, Counselor Stroupe generously provided invaluable reviews and advice throughout development of this manuscript.

Special acknowledgements to Mark Thinnes, Seattle, Washington, who designed the cover, and to Cindy Harris, Pleasanton, California, who photographed the image of "Little Joe" and his Dad that appears on the cover.

A special acknowledgement to John Farrington, past Principal of South Junior High School, Colorado Springs, Colorado, for whom I had the honor of working. The ideal Principal, he put the students and their needs first, because they are our hope for tomorrow. John led by example.

CONTENTS

CHAPTER 1
Crisis

***Great emergencies in crises show us how much greater
our vital resources are than we had supposed.***
William James[1]

*On September 11, 2001, forever to be known as "9-11," the single
worst act of terrorism was committed on the soil of the United States of
America.* On that infamous day, more than 3,000 Americans died when
two hijacked jetliners crashed into New York's World Trade Center,
causing the twin towers to collapse. Almost immediately a third jetliner
smashed into the Pentagon, and a fourth hijacked plane crashed in
western Pennsylvania — *All at the hands of Islamic terrorists.*

8:46:40 — American Airlines Flight 11, enroute from Boston (Logan)
to Los Angeles with 92 souls on board, crashed into the North
Tower of the World Trade Center.

9:03:11 — United Airlines Flight 175, enroute from Boston (Logan)
to Los Angeles with 65 souls on board, crashed into the South
Tower of the World Trade Center.

9:37:45 — American Airlines Flight 77, enroute from Washington, DC (Dulles) to Los Angeles with 64 souls on board, crashed into the Pentagon.

9:58:59 — South Tower collapsed.

10:03:11 — United Airlines Flight 93, enroute from Newark to San Francisco with 44 souls on board, crashed near Shanksville, Pennsylvania (60 miles southeast of Pittsburgh).

10:28:25 — North Tower collapsed.

In an elapsed time of one hour and 15 minutes, the number of dead totaled 2,992. This sum included 343 New York City (NYC) firemen, 23 NYC policemen, 37 Port Authority police, 3 emergency medical personnel (EMP) who perished at the scene, and 406 first responders subsequently died.

Our children, and their children, will never know the peace, security, and optimism that our parents and we have known.

On "9-11" America changed —
never again will it be the same.

Seven years have passed since that fateful day. The time has come for leadership to step up, take charge, and move this country in a new direction. It is no longer business as usual. As far as energy policy is concerned, it is not movement toward isolationism that is needed — it is movement toward energy independence. As Carl Gustav Jung, Swiss psychiatrist, 1875-1961, said, "We cannot change anything unless we accept it. Condemnation does not liberate, it oppresses."[2]

Study after study has been conducted by Department of Energy (DOE) Federal Laboratories and their myriad of contractors, at a cost of untold, and probably uncountable, billions of dollars, to say nothing of the manpower expended on the study of issues associated with nuclear power, only to be shelved without any action being taken for the good of the country or for its people. This lack of effectiveness was a matter of concern in the Administration and the Congress in the mid-1990s.

Former Vice President Al Gore's National Performance Review (1993) reported inefficiencies as high as 40% within DOE's Environmental Management program that could lead to losses of more than $70 billion over the next 30 years.[3] Victor Rezendes, then Director, Energy and Science Issues, Government Accounting Office, in Congressional hearings (1995) testified that DOE suffers from serious management problems, ranging from poor environmental management to major internal inefficiencies, and that poor oversight is the cause of the problem.[4] The Agency's continual efforts to re-align itself and justify its existence are the root cause of its mismanagement. Senator Rod Grams (R-MN) and Representative Todd Tiahrt (R-KS) proposed legislation that would downgrade the department to an agency, privatize U.S. oil reserves and DOE laboratories, set new guidelines for toxic waste disposal, and transfer control of the nuclear weapons stockpile to the Department of Defense (DoD).[5] But their initiative failed.

No progress has been made in the U.S. commercial nuclear industry since the Three Mile Island (TMI) Nuclear Generating Station's pressurized water reactor accident on March 28, 1979.[6] There were no deaths or injuries to plant workers or the 25,000 members of the nearby community on the Susquehanna River near Harrisburg, Pennsylvania. The accident involved a partial meltdown of the core in TMI-Unit-2; its core has been removed, but Unit-1, which came on-line in April 19, 1974, continues to operate today. It is licensed to operate through April 19, 2014.[7] The accident led to serious economic and public relations consequences that destroyed the public's confidence in nuclear power.

Seven years later, Russia's Chernobyl accident caused civilian casualties, but it did not deter Russia's pursuit of nuclear energy. (The eventual toll is unknown, but a 2005 study of the accident concluded, "The nuclear disaster at Chernobyl almost 20 years ago has so far claimed fewer than 50 lives."[8]) Other countries in Europe and the Far East have been more pragmatic, and they have progressed, most notably France. Meanwhile, the United States has stood still. Seven years after 9-11, the following represent the de facto situation:

- **Given:** That the United States is in a state of war against radical Islamic terrorism and remains dependent on oil imported from Middle Eastern States sympathetic to terrorists;

- **Given:** That the United States has been unable to resolve pervasive, political problems associated with long-term storage of spent nuclear fuel and surplus nuclear materials;

- **Given:** That state-of-the-art nuclear power generating systems using recycled nuclear materials are available;

- **Given:** That nuclear fuel is available from spent nuclear fuel and deactivated nuclear weapons systems;

- **Given:** That the current national energy policies, based on valid scientific and economic analyses, acknowledge the advantages of nuclear power, but remain inert because of the emotional issues raised by TMI;

- **Given:** That there is a critical shortage of electrical power in the United States because of inadequate generation capacity exacerbated by inefficient distribution networks.

Temporary storage of spent fuel has become an increasingly difficult problem for U.S. policy makers because permanent disposition solutions are so politically charged by environmentalists that agreement amongst stakeholders has proved impossible. As of October 2007, the United

States had generated approximately 63,000 metric tons of heavy metal (MTHM) of spent nuclear fuel from commercial nuclear power plants, and this amount could double by 2035. Ninety-eight percent of the 47,000 MTHM is stored in 33 states with 90% at nuclear plant sites. Additionally, there will be about 7,000 MTHM of spent fuel from research reactors, U.S. Navy reactors, prototypes, and those reactors used to produce nuclear weapons materials.[9]

The majority of this spent fuel is in temporary storage at DOE sites in Washington (Pacific Northwest National Laboratory [PNNL], Richland); Idaho (Idaho National Laboratory [INL], Idaho Falls); Colorado (Fort St. Vrain Nuclear Generating Station [FSV], Platteville); South Carolina (Savannah River National Laboratory [SRNL], Aiken); and Tennessee (Oak Ridge National Laboratory [ORNL], Oak Ridge).[10] Additional nuclear materials (surplus plutonium [Pu][11]) will come from decommissioned nuclear weapons.[12] DOE is supposed to take ownership of all nuclear fuel if and when Yucca Mountain becomes operational. This spent fuel can be a source for plutonium and uranium to use in manufacturing of nuclear fuel.

Since TMI, development of new U.S. nuclear power generation capability has been at a standstill. Meanwhile, the demand for electricity has continued to grow, and the power deficit, particularly in the major metropolitan areas of the United States, increased in magnitude. History shows that reliance on oil imported from politically unstable areas is a high-risk energy policy. The Organization of Petroleum Exporting Countries (OPEC) oil embargo of the early 1970s hammered the U.S. economy, increased inflation, and decreased industrial growth. At that time, the United Stated imported 35% of its oil – today it imports more than 66%, and, if the trend is permitted to continue, imports could climb to 85% by the year 2030.[13] If fuel consumption increases at this rate, then imports have to grow. Should rogue, anti-American nations operating worldwide succeed in cutting off supplies, the impact on the United States would be devastating.

Electrical power blackouts on the West Coast that resulted from insufficient power generation capacity was considered one of the nation's most important events of 2001, following closely the terrorist attacks of September 11, anthrax scares, and U.S. involvement in the war in Afghanistan that began less than a month later in response to the 9-11 attacks.[14] Then came our involvement in the Iraq war in 2003, which developed into a civil war amongst the Shiites, Sunnis, and Kurds.[15] All these are good and sufficient reasons for the United States to redirect its national energy policy to support nuclear power to generate electricity, thereby, contributing to solving or alleviating six major national issues:

1. Dependence on foreign sources of oil to meet its energy needs;

2. Long-term storage of spent nuclear fuel and nuclear components of deactivated weapons systems;

3. Proliferation of nuclear weapons of mass destruction, particularly by rogue nations, and the potential use by terrorists of a "dirty" bomb;

4. Shortage of electricity generation capacity, especially in the Northeast and West Coast;

5. Adverse impacts on the environment, such as global warming, air pollution, and depletion of natural resources by coal- and gas-fired electrical power plants.

6. Nuclear power as the foundation for a hydrogen economy.

The best way to resolve these issues is to pursue aggressively nuclear power generation using recycled spent nuclear materials as reactor fuel. Replication of France's proven nuclear fuel recycling capability is the most expeditious, effective, and efficient course of action to resolve the above-specified national issues. France, the world leader, has mastered the entire nuclear fuel cycle, from the extraction of natural uranium (U), to enrichment and manufacturing techniques, to recycling of spent fuel. By standardizing the reactors, France made it possible to link, expeditiously and effectively, those involved in research and development (R&D), industry, and the oversight and safety authorities.

The leadership and expertise that France maintains in its complete cycle solutions to nuclear power application in the generation of electrical power are commendable. The United States should see the potential advantages to be gained by replicating these "best practices" gleaned from the French experience that are applicable to production of electricity in the United States. This can be facilitated by utilizing AREVA COGEMA to close the gap in U.S. technology and infrastructure.

Further, and again, as a result of 9-11, we find our people and their outlooks have changed, and that more realistic and pragmatic approaches must be taken toward solution of national problems.[16] Thus, there is a unique window of opportunity to revitalize our nuclear resources in general and, in particular, to build our capability to generate electricity by utilizing modern, state-of-the-art reactors that are safe, efficient, and cost-effective. The problem is that small special interest groups effectively shut down further development of nuclear power generation in the United States after TMI in 1979. Meanwhile the rest of the world continued to pursue its nuclear programs with vigor.[17]

Because of the continuing threat of radical Islamic terrorism and the increasingly high cost of foreign oil, our national priorities have to change. The time has come to re-examine and revise our national energy policies and priorities and to take more timely and pragmatic approaches toward solution of issues that adversely affect the best interests of the United States. Pakistan and India stay mobilized along their common border in a nuclear standoff.[18] Three wars have been waged since Pakistan gained independence from Britain in 1947. These two countries have a deep seeded, religious hatred of one another. It stems from those who embrace Hindu nationalism and those who embrace Islamic fundamentalism. India tested its first nuclear bomb in 1974. Then, in 1989 Pakistan tested its first nuclear bomb.[19] Rogue nations, Iran, Libya, Syria, and North Korea, are particularly worrisome.

Currently, the most pressing national crisis is the involvement of U.S. forces in Afghanistan and Iraq, and ultimately the resolution of what, in Iraq, has become a civil war. As a result, the United States should resolutely pursue courses of action to reduce its dependence upon foreign oil and its long-term nuclear waste storage problems. Spent nuclear fuels can be recycled to provide fuel for current or state of-the-art reactors to generate electricity. (Europe runs some existing nuclear plants with 20% of the fuel from mixed oxide (MOX), which is made with recycled fuel. MOX is discussed in Chapter 2.) A significant byproduct of this effort is that it provides a rational and cost-effective solution to the nuclear waste problem — a solution that rids the major nuclear powers of their excess weapons-grade plutonium and uranium.

Terrorists can easily obtain radioactive materials to make "dirty" bombs by simply attaching radioactive nuclear material to a conventional bomb or other explosive device. There is no nuclear explosion; the nuclear material is simply spread by the explosion of the conventional weapon. Speculation as to the credibility of this threat has gained momentum recently, because missing inventory of nuclear devices and materials in the United States and the former Union of Soviet Socialist Republic (USSR) have not been located.[20] There are reports emanating from Afghanistan that nuclear materials have been found in a large tunnel complex once controlled by Osama bin Laden's al-Qaeda terrorist network.[21] These bombs would not necessarily cause *mass destruction,* but they could cause *untold casualties and major disruption of our infrastructure and economy.*

A tiny amount of radioactive material in the wrong hands could be devastating; "dirty" bombs in the hands of terrorists pose a real and deadly threat. The potential impact of a "dirty" bomb can be appreciated by considering the 1987 incident that occurred in Goiania, Brazil. There, scrap scavengers broke into a radiological clinic and stole a capsule containing about an ounce of cesium 137. The highly radioactive material was cut into more than 100 pieces and passed along to friends and family. "Fourteen people were exposed to radiation out of 249 contaminated. Four subsequently died, and more than 110,000 had to be continuously monitored. To decontaminate the area, 125,000 drums and 1,470 boxes were filled with contaminated clothing, furniture, dirt, and other materials; 85 houses had to be destroyed."[22]

Another example of the effects of radioactive material occurred in November 2006 when a former KGB agent and Putin critic, Alexander Litvinenko, died in London from Polonium210 poisoning. He suffered extensively for 22 days before dying, blaming the Russian authorities, and specifically naming President Vladimir Putin as the person responsible. In the process of being smuggled into the United Kingdom (U.K.), the nuclear material exposed the passengers on the British Airways (BA) aircraft, and later the Londoners in the vicinity of the restaurant where Litvinenko was poisoned. Also, as the material was transported through the streets of London, many additional citizens were exposed. Although Russia was going for one of its own, thousands of British citizens had been affected. Ironically, the agent who is supposed to have committed the murder is now running for public office in Russia.[23]

Addressing the threat of nuclear terrorism in the 2004 presidential campaign, both President George W. Bush and Senator John Kerry called that threat "the single greatest danger facing the American people." Al-Qaeda leaders have named seven American cities targeted for nuclear destruction: New York, Washington DC, Miami, Chicago, Houston, Las Vegas, and Los Angeles.[24] And, at a conference in Tehran in October 2005, Iran's President Ahmadinejad called for the destruction of Israel; specifically, he said Israel should be "wiped off the map."[25] Later, he called for the destruction of the United States as well. What will nuclear-minded Iran do? We do know from experience in Iraq that Iran *will* provide weapons to terrorists and others engaged in hostile activities against Western nations.

Terrorist cells are active worldwide, including the United States, and they have carried out attacks on nearly every continent. Of major concern are sleeper cells associated with Islamic terrorists groups that have been located in 40 states, including North Carolina, South Carolina, New York, New Jersey, Michigan, Oregon, Virginia, West Virginia, Georgia, Alabama, Texas, Oklahoma, Florida, Pennsylvania, California, Illinois, Ohio, Washington, Colorado, Tennessee, Vermont, Arizona, and in America's backyard, Montreal and Ontario, Canada.[26] They are positioned strategically, especially in rural areas. Of special interest as potential targets are subway systems, air ducts in large office complexes, sports arenas, schools, hospitals, malls and shopping centers, airports, water treatment facilities, electrical utility infrastructures, chemical storage facilities, research facilities, important landmarks, hospitals, critical components and sections of highway and railroad systems, and military bases. Just how many sleeper cells and exactly where they are located we don't know, but we are well aware of their objectives.

Usually, terrorist cells are composed of four to 12 individuals, not including the support cells, whose purpose is carrying out attacks of violence.[27] (The al Qaeda cell that carried out the 9-11 attacks, for instance, subdivided into four-man teams, each assigned to a specific aircraft.) The terrorist cells are ready to strike when materials and timing coincide with target opportunity. The list of possible materials of death is comprehensive and frightening. In addition to ugly, often horrific, "conventional" devices, such as car and truck bombs; roadside bombs (often called IEDs [improvised explosive device]); human "bombs" (first men, then women, sometimes children, and recently even the handicapped) on suicidal missions. The list also includes biological, chemical, and nuclear materials — potential methods of employment are limited only by one's imagination. We do know, for example, that bin Laden's al Qaeda potential nuclear capabilities played a role in the third general alert, issued by then director of Homeland Security Tom Ridge on December 3, 2001.[28] We don't know where or when, but we can be certain that terrorists will strike again.

If the United States generates a market for spent fuel, we would be in control of its final destiny and accountability.[29] Figure 1 is a roadmap to be followed to reach that objective. Tracking from left to right, following the bold-line path, nuclear materials taken from two sources — spent fuel and deactivated warheads — should be recycled and fabricated into fuel. The electrical power generated will then be distributed through the grid to the consumer. The nuclear waste remaining (10%) would be either prepared for long-term storage or recycled. Multiple recycling in advanced reactors offers significant advantages, not only in lowering costs, but more importantly, in reducing the amount of nuclear waste.

Relating to nuclear power, important questions to be answered are the following:

1. Is technology available to support such an alternative?
2. Does the United States have the capability to utilize available technologies to implement the alternative?
3. Does the alternative source of power help solve one or more major national issues?
4. Can the alternative be implemented in the near future?
5. Is the implementation of the alternative economically feasible?
6. Can the alternative be sustained through the long term?
7. Will the alternative provide sufficiently large amounts of electrical power reliably and safely in an environmentally sound system?
8. Finally, is there, then, a viable alternative to long-term storage of nuclear waste?

The nuclear power industry in France that generates 80% of its electricity and the U.S. Navy Nuclear Propulsion Program (NNPP) that powers its aircraft carriers and submarines, have demonstrated long-term, effective, efficient, and safe operation of nuclear power plants. *Serious effort must be forthcoming toward the preservation of our country, especially for the children. Their future is in our hand. Nuclear power can be a major contributor to this goal.*

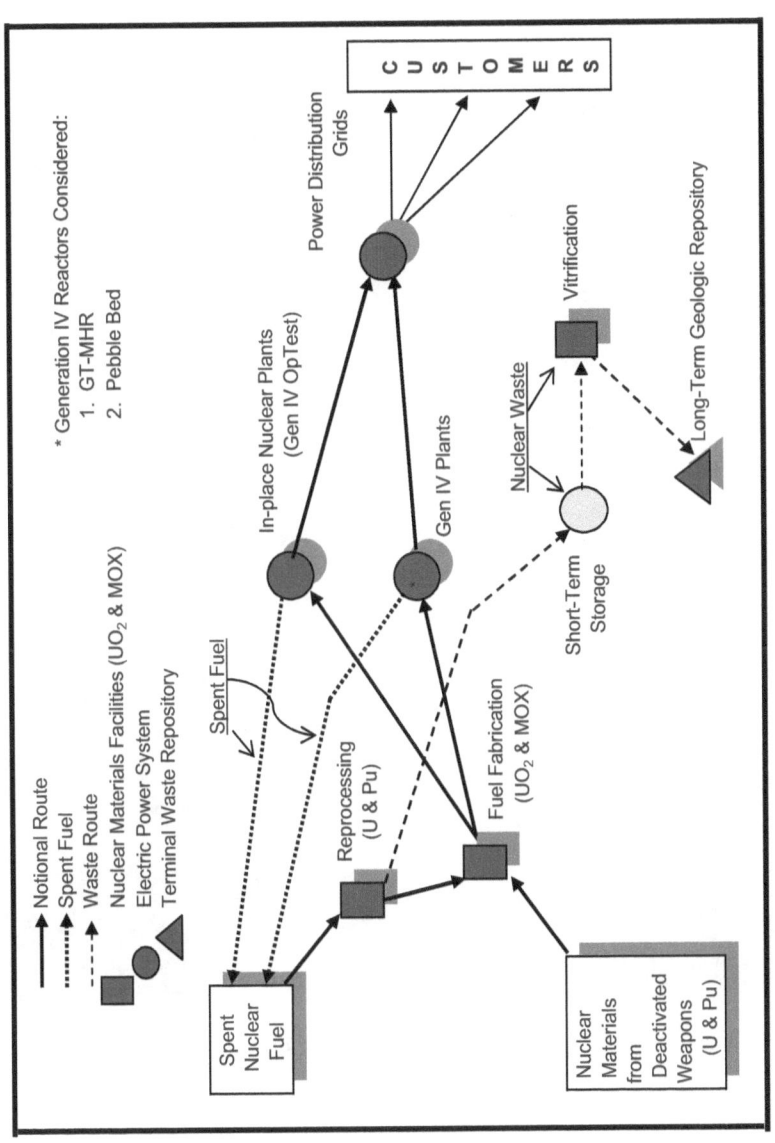

Figure 1 Roadmap

CHAPTER 2

Nuclear Issues

Knowledge is power.

Sir Francis Bacon[30]

The late Dr. Edward Teller, a renowned physicist and prominent figure in the second half of the 20th century, called the development of the hydrogen bomb or "super bomb" an important piece of scientific history. He was often referred to as the father of the hydrogen bomb, an expression he disliked. During the Reagan years, he frequently was called upon to advise the President on the Star Wars program.[31]

In the Christian Science Monitor, Teller wrote, "the nuclear-power industry in the United States had the technical capability to make reactors that can't melt down, but it just hasn't done it." He identified two reasons: lack of public support and high cost. As a result, 70 years after the beginning of the fission age, nuclear technology is still struggling with acceptance, and the United States is still dependent on oil and natural gas imported from Organization of Petroleum Exporting Countries (OPEC). Times have not changed. Teller asserted that the old systems are too large, too complicated, and too onerous. They incur high operating costs that are exacerbated by spiraling regulatory requirements.[32] He, too, was concerned about radioactive wastes with half-lives of thousands of years and with nuclear weapons proliferation in emerging nuclear nations. New reactor designs can solve this problem.

The growing world energy requirements, especially in third world countries, must be dealt with or else they will result in consuming huge amounts of irreplaceable fossil fuels, ripping away forests, or fighting for resources to sustain themselves. Teller said that, in his opinion, the best solution is with a truly modern, inherently safe reactor, and one in which the public would have confidence. His interest piqued by the discovery of the natural underground reactors at Oklo, in Gabon, Africa, he came to believe that underground siting was the ultimate answer to the problem of nuclear safety. In his last paper, "Thorium fueled underground power plant based on molten salt technology," Teller advocated a shallow underground setting for the Liquid Fluoride Reactor (LFTR). The paper was written in collaboration with Dr. Ralph Moir (retired research physicist, Lawrence Livermore National Laboratories [LLNL]).[33] Teller was not alone in holding this idea, Andrie Sakharov (Russian nuclear physicist, Nobel Peace Prize 1975) wrote in his Memoirs, "Plainly, mankind cannot renounce nuclear power, so we must find technical means to guarantee its absolute safety and exclude the possibility of another Chernobyl. The solution I favor would be to build reactors underground, deep enough so that even a worst case accident would not discharge radioactive substances into the atmosphere."[34] Siting reactors underground should be an objective to enhance safety and security, as in the Notional Nuclear Power System (NNPS) proposed in Chapter 10. To facilitate achievement of this objective, reactors should be built in factories with cost and quality control regulations in place. The reactors would be relatively economical to acquire and cost-effective to maintain.

Reactors operating today consume 50% more nuclear fuel, create 50% more waste, and exhaust 100% more thermal energy to the environment than would the advanced (Generation IV) reactors using the latest technology. Marjorie Mazel Hecht (Science Editor, Executive Intelligence Review) reported her views on two new reactor designs that, given the proper political resolve, could be on-line within this decade.[35] They are modular, mass-producible, easy to operate, super-safe, and affordable.

Americans need to be aware of the contributions nuclear has made and can make to their environment. Teller said that even the Club of Rome believed the world is headed for serious energy problems because of its reliance on fossil fuels. (The Club of Rome is a global think tank and center of innovation that brings together scientists, economists, international high civil servants, heads of state, and others dedicated to the improvement of society. Its mission is to act as a global catalyst of change that is free from any political, ideological, or business interest.)[36]

Currently, the United States has 104 commercial operating nuclear reactors in 31 states that supply about 20% (in 2007, 807 billion kilowatt-hours) of the nation's electricity, safely, cleanly, and efficiently (91.8% capacity factor at a cost of 1.68 cents per kWh, the lowest production cost of any major source of electricity).[37] Nuclear power plants are located in Alabama, Arizona, Arkansas, California, Connecticut, Florida, Georgia, Illinois, Iowa, Kansas, Louisiana, Maryland, Massachusetts, Michigan, Minnesota, Mississippi, Missouri, Nebraska, New Hampshire, New Jersey, New York, North Carolina, Ohio, Pennsylvania, South Carolina, Tennessee, Texas, Vermont, Virginia, Washington, and Wisconsin. The 69 Pressurized Water

Reactors on-line produce 65,100 megawatts of electricity, and the 35 Boiling Water Reactors produce 32,300 megawatts. The ten top states in percentage of nuclear power usage are Vermont 72.5, South Carolina 52.0, New Jersey 50.6, Illinois 48.0, Connecticut 46.6, New Hampshire 39.2, Virginia 35.4, Pennsylvania 35.0, North Carolina 30.8, and New York 29.0.[38]

No new U.S. reactor had been ordered since the middle 1970s, primarily because of peoples' mindset against nuclear power and the failure of public leaders, for political reasons, to support nuclear expansion or initiatives. Construction of Watt's Bar Unit-1 began in 1973, prior to the TMI accident, but was halted during the 1980s. Completed and approved in 1995, it was put on line in 1996. Finally, after a hiatus lasting nearly three decades, Reuters (2007) reported that NRG Energy, Inc. is filing an application with regulators to build two new nuclear reactors at its facility in Bay City, near Houston.[39] The two 1,350-megawatt General Electric advanced boiling water reactors (ABWR) would be operational by 2020. The cost of the project is estimated to be $6 billion, create 4- to 6,000 temporary jobs, and is hailed by Matagorda County Judge Nate McDonald as *the beginning of the nuclear renaissance in America.*

In an article titled, "Train Wreck Stirs Fear Over Nuclear Freight," Matthew Mosk (2001) wrote, "Efforts to exploit situations like the Baltimore fire point up the fact that *it's a political debate, not a technical one* (emphasis added)."[40] This was in reference to the fact that there have been no nuclear initiatives since Three Mile Island (TMI). There was a massive public relations campaign funded by oligarchic interests, "environmental" groups, and huge U.S. foundations that scared the public regarding nuclear power.

The Carter Administration was run by a Trilateral Commission that was determined to establish a post-industrial economy. It thwarted long-term investment with interest rates exceeding 20% and unreasonable, time-consuming regulations and restrictions on new construction of nuclear power plants. The environmentalists established a stronghold in Washington during Carter's Presidency. They accomplished virtual paralyses of new Federal coal leasing, conventional electric generating plant licensing, mineral land leasing and water development, and halting of new nuclear power construction. The nuclear industry refused to fight back. In retrospect, it is ironical indeed that President Carter, a nuclear-trained Navy officer, generally took negative positions on nuclear power and allowed the program to languish from lack of leadership and nourishment. *The Carter administration made reprocessing of nuclear fuel illegal.*

Over the past 25 years, instead of building the 2,000 nuclear power plants by the year 2000, as envisioned by the Atoms for Peace program in the late 1950s, the world has only 440 operating nuclear plants. And, the United States is in a huge energy crisis, as evidenced by the breakdown of the utilities in California in the spring of 2001 "... that seemed to become more critical" with each passing day. As with any crisis, comes an opportunity — an opportunity to rejuvenate the U.S. nuclear power industry. But, why use Nuclear Power? There are several reasons:

- U.S. dependency on foreign oil has had a stifling effect on the economy.
- The energy produced per amount of material consumed is by far the highest of   the major sources — coal and gas.
- The costs are competitive with coal, the major energy source in the world.
- Compared with alternative energy sources, the impact on the environment by nuclear is minimal. No greenhouse

gasses are emitted, as with coal and oil. And, no large land areas are required, as with solar and wind.

- Uranium has virtually no other use and is abundantly available.

- Plutonium, a by-product of commercial nuclear plant operations, also can be used as fuel, and it too is readily available.

- The amount of waste produced is the least of any major energy production process (although a final disposition has not been mandated).

- Nuclear energy provides benefits other than generation of electricity, in particular, diagnostics and therapeutic treatments in medicine, radiography, irradiation of food and mail, weld inspection, and power in remote geographic locations, including the Antarctic and space.

The American nuclear industry is its own worst enemy. Costly projects have been stalled for political and regulatory reasons. Utilities should be focusing on an innovative and inherently safe reactor design that is environmentally sound and economically competitive. This reactor type should use new fuels, new high-efficiency designs to reduce heat emission or transfer to the environment, smaller units that can be built and tested off-site, and self-contained units that can be installed in secure, out-of-sight, underground facilities that can be located proximate to the electrical grids where most needed. Engineers and scientists worldwide are excited about these new reactors, but industry and government leaders in the United States are pessimistic about the public's willingness to accept nuclear power under any circumstances.

In the end, the future of nuclear power rests in the hands of public opinion. As long as people remember Chernobyl and TMI there will be a mistrust of the nuclear industry and the government agencies that regulate it. But, if, as a result of the attacks against America on September 11, 2001, and the Iraq and Afghanistan Wars, the nuclear debate "resumes its old intensity," it will be based on a new set of arguments.[41] This intolerance is wrong. *It is time to educate the public and take a closer look at modular reactors.*[42]

Oil spillage, waste materials that last forever, global warming, polluted beaches, and steadily increasing atmospheric pollution from burning fossil fuels are persuading many political leaders to review these problems. Americans who want a clean, safe, and domestically produced energy source should embrace nuclear power. In 2005, Senator Pete Domenici sponsored a bipartisan nuclear energy bill that fostered greater use of nuclear energy and supported advanced research into technologies to minimize waste.[43] Where is this initiative now?

The United States has been buying highly enriched uranium from Russia, primarily because of its fear of diversion of this material to the making of terrorist bombs. Since 1991, the United States has spent about $7 billion on Russian security to help Russia meet its arms-control treaty commitments.[44] It also has paid to slice hundreds of nuclear-launch missiles, submarines, and bombers into scrap metal. The U. S. non-proliferation policies and terrorist fears have not been stressed in the U.S. nuclear energy policy.

With the onset of the "Global War on Terror" and establishment of the U.S. Department of Homeland Security, two national issues arose into the limelight of public consciousness — first is America's dependence on imported oil, primarily from the Middle East, an area seemingly always and forever in turmoil; the second is the long-term disposal of nuclear waste, both of which are of grave concern to our leaders — the keepers of the peace. The United States faces huge challenges in the years ahead.

According to then-Ambassador Richard Holbrooke, "Our greatest single failure over the last 25 years was our failure to reduce our dependence on foreign oil…which would have reduced the leverage of Saudi Arabia."[45] Further, James Woolsey said that Saudi Arabia believes the United States can be controlled because of its dependence on Saudi's oil.[46] (Imported oil accounts for about two-thirds of U.S. consumption.) The sources of oil for the first six months of 2008, January through June, are as follows: Canada 21%; Saudi Arabia 16%; Mexico 13%; Nigeria 12%; Venezuela 11%; and Others 27%.[47] The Organization of Petroleum Exporting Countries (OPEC), a "cartel" with arbitrary power to control the quantity and price of oil that reaches the marketplace, controls about two-thirds of the world's oil reserves and one-third of its production. It was the supplier of 60% of U.S. oil imports during the period.[48] The Center for Strategic and International Studies predicts that the United States will grow steadily more dependent on oil imports, which will exceed 75% of its consumption by the year 2020.[49]

Vice President Cheney stated that America's dependence on oil from overseas threatens national security.[50] Over-dependence on any one energy source, especially a foreign source, leaves the country vulnerable to price shocks, supply interruptions, and, in the worst case, blackmail. Oil prices reached a then-record high topping $100 dollars a barrel in September of 2007; six months later the price of a barrel was more than $110, and the price peaked at $147 a barrel in July of 2008.[51] Large reduction in use of foreign oil requires a new fuel for our cars — hydrogen may be the best.[52] If action isn't taken to remedy this situation, those imports are expected to grow. Refining capacity, now running at 96%, must be increased. Any alternative source of energy that lessens the demand for oil should be considered as a matter of priority; however, of all the sources of energy being pursued, the most viable, environmentally friendly, cost-effective solution is nuclear energy. It has the greatest potential. The view that nuclear power should be categorically opposed without regard to our energy needs, our dependence on countries that harbor terrorists, foreign dictators, our quality of life, or any other ramification is unbelievable. Nuclear power can and should be harnessed in a safe and prudent manner.

The legacy of the arms race has presented the DoD and DOE with enormous tasks to clean up nuclear waste. The DOE must dispose of 50 tons of weapons-grade plutonium, and it must pursue an approach that will render this plutonium inaccessible and unattractive for future weapons applications. It also needs a program for long-term isolation of unwanted, intensely radioactive fission products in spent fuel; indefinite storage of spent fuel at nuclear power plants or some temporary, central location, is not the solution. Plutonium, uranium, and valuable fission products present in spent fuel should be removed prior to transfer of the waste to a permanent repository, such as Yucca Mountain, Nevada. These materials are *valuable*, and that value should be realized for the benefit of society rather than discarded. Long-lived fission products can be burned in appropriately designed reactors.

Because of the highly charged state of affairs in the United States regarding the disposition of nuclear waste, a debate that is primarily political, it is clearly in the best interests of the United States that we design (presently underway), construct, and operate MOX fuel fabrication facilities to convert excess weapons-grade plutonium into usable fuel for nuclear reactors. (MOX is mixed oxide fuel: $PuO2$ [plutonium oxide] is mixed with $UO2$ [uranium oxide]. The mix is generally 5% $PuO2$.) The resulting product could then be utilized productively as fuel in modern reactors to produce electricity — solving, at least in part, these two major national issues: imported oil and nuclear waste.

Dr. James A. Lake, past president of the American Nuclear Society, said that the volume of high-level nuclear waste generated by America's 104 nuclear plants during their lifetimes could be stacked in a space the size of a football field to a depth of approximately 15 feet.[53] By comparison, the equivalent number of coal-fired plants produces thousands of times more waste by volume, and its toxicity lasts forever. According to John Stuckless, by 2001, DOE had spent more than $6 billion to research geologic, engineering, and transportation issues associated with Yucca Mountain.[54] (By 2005, that figure had reached $12 billion, and it is expected to reach $23 billion by the revised Yucca Mountain opening date of 2017.[55]) His book includes discussion of significant findings of DOE's Site Characterization Study, which contained results of more than 20 years of scientific research and analyses by hundreds of scientists. The nuclear power utilities have committed $28 billion (of rate payers' money, not tax money) to the nuclear waste fund to pay for the permanent disposal of their spent nuclear fuel (SNF) in a geologic depository.[56] The fund also is used to help defray the costs of developing and operating the repository in the post-emplacement years.

Dr. Lake also suggests that the United States could learn something from France, Britain, and Japan where nuclear energy is a major component of electricity generation. "These countries have the technology and *political will* to recycle spent fuel rather than throwing it away after a single use. This strategy extracts more energy from the fuel, while at the same time reducing the volume and radioactive lifetime of the wastes that have to be stored in a repository."[57]

Although reuse of spent fuel imposes additional transportation requirements, transportation of nuclear materials has not been a problem. The U.S. nuclear energy industry has completed more than 3,000 shipments of used nuclear fuel over the past 40 years with no injuries, fatalities, or environmental damage as a result of the radioactive nature of the cargo.[58] *There has never been any accident in the United States where there was injury due to nuclear materials.* Tests have proven that the specialized rail cars used in transporting nuclear materials are as indestructible as bank vaults. "Promotional videos released by the industry show images of rail cars slamming at 100 mph into solid concrete walls, dropping from three stories to the hard earth, and being engulfed in flames accelerated by jet fuel," said Steve Unglesbee, spokesman for the Constellation Energy Group that operates the Calvert Cliffs Nuclear Power Plant.[59]

Russia is facing the nuclear waste disposal problem as well. And this is of concern, because questionable security and accountability practices extant in Russia could provide easy access to terrorists and rogue nations eager to get their hands on nuclear materials. According to the Nuclear Energy Minister in 1997, Viktor Mikhjailov, Russia had dismantled almost 50% of its nuclear arsenal. Russia's estimated 8,000 to 9,000 nuclear warheads would be reduced to no more than 3,500 under the START II treaty. Currently, (January 2008), Russia is estimated to have about 4,000 strategic nuclear warheads (Arms Control Association). The picture is fuzzy regarding the security and accountability of non-strategic or tactical nuclear weapons, which, broadly speaking, can be any nuclear weapon not classified as strategic, and they are not covered in SALT or START agreements. (This belittles the threat they pose, because even a relatively small nuclear warhead could destroy a city, and their "smallness" makes them attractive to terrorists – easy to steal, easy to hide, and easy to employ.) Responding to President George H. W. Bush's 1991 unilateral initiative to reduce tactical nuclear weapons and proposal that Russia follow suit, it is believed that Russia has reduced its tactical nuclear warheads by as many as 18,000. But the reductions have not been transparent; how many remain and where they are employed or stored are unknown. The potential threat continues to be of serious concern.

Although Russia has made significant reductions in its nuclear forces since the end of the Cold War, the pace of reductions was problematic because of an initial lack of funding. Also, Russia has assumed control of all nuclear weapons that were in the former Soviet republics, including strategic weapons deployed in Kazakhstan, Ukraine, and Belarus.

As a result of non-proliferation agreements between the United States and Russia, the United States declared 175 metric tons of highly enriched uranium (HEU) from the nuclear weapons program surplus in 1995.[60] The 77% that is "commercial grade" can be blended with U238 to produce low enriched uranium (LEU) for use in commercial nuclear reactors without further recycling. About 40 metric tons is not "commercial grade" and requires special processing before it can be utilized productively. The DOE issued a Disposition of Surplus Highly Enriched Uranium Final Environmental Impact Statement (FEIS) in 1996 that considered alternatives for disposition of the surplus HEU, which included the "off-specification" material.[61] The Record of Decision (ROD) was issued along with the FEIS stating that the surplus HEU would be made non-weapons-usable by down-blending it to LEU. To the extent practical, its economic value could be recovered by using the LEU as reactor fuel.

In 1997 DOE and the Tennessee Valley Authority (TVA) signed a memorandum of understanding (MOU) to pursue a joint program to investigate the use of 33 of the 40 metric tons of off-specification HEU as a source of LEU for TVA's reactors. The TVA solicited proposals from prospective bidders for converting the HEU to useable forms, down blending it into LEU, and processing the LEU into fuel assemblies. In 2001, TVA adopted DOE's FEIS and took steps to implement the preferred alternative course of action, which was to maximize commercial use.[62]

The world's net electricity consumption is expected to increase to 25 trillion kilowatt hours (kWh) by 2015, a 119.3% increase from the 1995 level of 11.4 trillion kWh. In 2004, electricity consumption reached 16.33 trillion kWh, a 43.2% increase from the 1995 level.[63] *To meet the demand, the equivalent of about 6,000 power plants having 300-megawatt capacity will have to be built.* The U.S. Energy Information Administration estimates that by 2030, 258 gigawatts (GW) of new capacity will be required.[64] This equates to 250-500 baseload power plants rated between ½ GW and 1 GW, and costs could total approximately $412 billion in 2005 dollars.

The previously mentioned electricity crisis in California, also known as the Western Energy Crisis of 2000 and 2001, occurred when a third of the state's total generating capacity of nuclear and fossil was off-line for maintenance. The unfortunate timing was critical, because it hit during the peak summer load conditions. Several plants, totaling 2,700 megawatts, had used up their annual pollution credits, so they could not restart without severe penalty. Normally, the state imports about 20% of its power needs, but a dry, hot summer reduced hydroelectric availability throughout the Northwest. Imported electricity from Arizona and Texas helped, along with local gas-fired plants, but the demand forced natural gas prices to double. As a result, wholesale electricity prices throughout the West soared to unprecedented levels, reaching $750 a megawatt.[65] Electric utilities experienced a quadrupling of wholesale prices from generators, but they had their own prices capped, and they suffered about $12 billion in losses over a six-month period. Forced to act, California stepped in to bail out the largest utilities and re-regulate the system.[66]

This devastating electricity crisis, it turns out, resulted from the "gaming" of a partial de-regulated California energy system by giant energy companies, principally Enron Corporation and Reliant Energy.[67] The energy crisis was characterized by a combination of extremely high prices and rolling blackouts. Price instability and spikes lasted from May 2000 to September 2001. Because of price controls, utility companies were paying more for electricity than they were allowed to charge customers. This forced the bankruptcy of Pacific Gas and Electric and the public bail out of Southern California Edison. In December 2001, following the bankruptcy of Enron, allegations were made that the energy prices had been manipulated by Enron, and in February 2002, the Federal Energy Regulatory Commission (FERC) began an investigation of Enron's involvement.[68] That winter, the Enron Tapes' scandal surfaced. Had there been an adequate electricity supply, the crisis could have been averted. The rest is history.

Although environmental groups continue to promote alternative sources, solar and wind farms just cannot meet the demand. As a result of the crisis, California is looking more favorably at nuclear-based technology. For the first time since 1985, the average cost of generating electricity in nuclear power plants in the United States has dropped below that of coal-fired plants. Over the past two years, nuclear increased its productivity and currently is operating at more than 90% capacity.[69] Estimates indicate that over the next 20 years, oil consumption will increase by 33%, natural gas by well over 50%, and demand for electricity will rise by 45%. If America's energy production grows at the same rate as it did in the 1990s, we will face an ever-increasing gap.

Having experienced the Western Energy Crisis and lesser crises in other parts of the United States, especially the densely populated areas in the Northeast, such as New York City, Long Island, and Boston, and potentially facing even more serious shortages in the future, it is clear that new energy policies must be forthcoming from the Congress. Bold steps must be taken, even if they are not universally popular with certain elements of the public. California's rolling blackouts, the insolvency of the state's major utilities, and the rapid deterioration of the state's economy and finances, all are still fresh in our minds, even though they may not seem quite so onerous today as they did at the time.

Early in the Bush Administration, the White House had estimated that the nation must build between 1,300 and 1,900 new power plants in the next 20 years, and invest $150 billion just in new pipelines and transmission facilities. Credit Suisse First Boston estimated that the investment shortfall in power plants alone amounts to $1 trillion. The key steps that Vice President Dick Cheney and his energy taskforce recommended to President Bush and his Cabinet cited NUCLEAR: Encourage the Nuclear Regulatory Commission (NRC) to re-license existing plants that meet safety standards; provide a "deep geologic repository" for nuclear waste; work with allies abroad to develop reprocessing and fuel-treatment technologies. As of 2006, the nation's nuclear power industry is pursuing plans to build 14 new plants in 10 states over the next 20 years. The potential locations are Texas, Virginia, Alabama, Mississippi, Louisiana, Georgia, North Carolina, Florida, South Carolina, Illinois, Maryland, or New York.[70]

The legitimate market for international nuclear waste, spent fuel and the plutonium and uranium from de-activated weapons, created by the reuse of these materials to fuel reactors generating electricity, is a good thing. It would essentially dry up part of those sources for the black market trade in nuclear materials, not to mention a huge boost to the economy.

In a House Energy and Commerce subcommittee hearing on energy policy and air quality, it was noted that there is a resurgence of interest in nuclear power. The recent energy crisis led to the conclusion that natural gas and coal should not be the only fuel sources for developing future generating capacity. Coal, which fuels the majority of U.S. power plants, dirties the air. Virginia, Connecticut, and several other eastern states have brought law suits against owners of coal-fired plants in the Midwest seeking damages from acid rain. If successful, the costs will further advantage nuclear plants. Power plants fueled by oil and natural gas have become more expensive, and both have environmental issues as well. Tauzin, former Louisiana Congressman, told the House that federal regulators or Congress should address the following issues to make nuclear power more viable. They are as follows:

- Require NRC to administer its rules *in a consistent and evenhanded manner that does not discourage companies from future investment.*
- Prepare the commission to renew reactor licenses that are set to expire in a few years. The NRC had renewed licenses to extend the lives of only five reactors.[71]
- Work harder to solve the nuclear waste issue, because *it is not safe to store spent nuclear fuel in dozens of locations across the country.*
- Make it easier for nuclear power generation to remain a viable component of the national energy mix.
- Reauthorize the compensation and liability provisions of the Price-Anderson Act that expired in August 2002.[72]

It was concluded that, without the above measures, the industry probably would not construct or operate new nuclear facilities. Vice President Cheney echoed those remarks when he said that nuclear power could help alleviate concerns about global warming.

On January 23, 2002, then Energy Secretary Spencer Abraham announced that the Bush Administration would go ahead with a multibillion dollar program to dispose of 34 metric tons of plutonium taken from dismantled U.S. nuclear weapons under an agreement that calls for Russia to get rid of a similar amount. The Bush National Security Council (NSC) delayed the program that originated in September of 2000 by the Clinton Administration until it completed its review of growing costs and other related issues. Abraham said that all the surplus weapons-grade plutonium in the program would be converted to MOX and used in nuclear reactors. He also eliminated Clinton's proposal for vitrifying two metric tons of the radioactive material in protective glass logs for long-term storage. Overall, the proposed plan would cost $3.8 billion over 20 years. Abraham concluded, "There is an increased urgency to move forward with the elimination of surplus weapons-grade material like plutonium, because of proven technologies to eliminate this material."[73]

The DOE evaluated different strategies to dispose of this material, implementation of which would be delegated to its Surplus Disposition Program. The Program plan is to convert approximately 37 tons of weapons-grade plutonium into fuel for commercial nuclear power plants operated by Duke Energy. The DOE selected Duke, Cogema, Stone & Webster (DCS) as the contractor to design, construct, and operate a facility to fabricate the MOX fuel. In March 2005, NRC issued a Construction Authorization (CA) to DCS for a MOX fuel fabricating facility (MFFF) at the Savannah River Site in South Carolina. In October 2006, DCS requested the corporate name for the CA for the MFFF be changed to Shaw AREVA MOX Services, LLC; it was approved in November. Meanwhile, in September, MOX Services had submitted a License Application (LA) to the NRC.[74]

In his article, Guy Gugliotta stated that deep in the Bush Administration's energy plan is a reference to an alternative approach to disposing of radioactive waste from nuclear power plants. "Reprocessing" the plan asserts "could help alleviate one of the major drawbacks to nuclear energy." Further, ". . . (it) will continue to discourage the accumulation of separated plutonium world-wide." The plan encourages further research into reprocessing, which makes the fuel to be burned to generate electricity.[75] Currently, only France, the United Kingdom, Russia, Japan, and India reprocess spent fuel, and only France, Belgium, Switzerland, Japan, and Germany burn the resulting finished plutonium oxide in nuclear plants. The article is incorrect when it states that no government agency or business has ever recycled nuclear waste for commercial use on U. S. soil. Prior to President Carter's renunciation of reprocessing and plutonium breeder research in a secret 1977 executive order, *reprocessing was the policy* of the U.S. government. The executive order, Presidential Directive 8, was declassified in 1994 and survives today as President Clinton's Presidential Decision Directive 13. *For reprocessing research to resume, the directive should be either rescinded or reinterpreted.* President Bush's National Energy Policy included the recommendation that the United States consider technologies to develop reprocessing and fuel treatment technologies that are cleaner, more efficient, less waste intensive and more proliferation-resistant."[76]

The Bush administration intended to move forward on a Clinton initiative to enlist Russia in a joint program to each convert 34 tons of surplus plutonium from nuclear weapons into MOX.[77] If the deal were closed and the United States makes MOX at Savannah River, Duke Power, a commercial utility, would burn it in two reactors near Charlotte, North Carolina. The Energy Department would then reimburse Duke for plant modification and sell them MOX at a subsidized price below what Duke would have to pay for enriched uranium fuel.

President Bush's Energy Plan, the Administration's 163-page energy proposal that was submitted to the Congress in May of 2001, included the following key recommendations favorable to the nuclear industry:[78]

- Provide $1.5 billion in tax incentives to facilitate the sale of nuclear power plants.
- Streamline nuclear plant licensing procedures and ask Congress to renew the Price-Anderson Act, which shields nuclear power plants from catastrophic liability costs.
- Re-examine nuclear fuel reprocessing technology, abandoned two decades ago, as a way to recycle nuclear waste.
- Direct the Energy Department and the Environmental Protection Agency (EPA) to assess the potential of nuclear energy to improve air quality.

Critics said that it proposed little to address gasoline prices or electricity shortages in the West. It tilts toward expanding the production and use of coal, gas, and nuclear energy.

The report was a stark assessment of the nation's problems. It stated that America "faces the most serious energy shortage since the oil embargoes of the 1970s." The 105-point blueprint was designed to solve the long-term gap between supply and demand, calling for new oil drilling on federal lands, new nuclear plants, and $10 billion in tax incentives for Americans to save energy.

The House GOP Energy Bill highlights the following:

- $33.5 billion in tax cuts and incentives over 10 year to encourage oil, gas, "clean" coal, nuclear energy production, and conservation measures.

- Funds nuclear energy research, including fuel reprocessing.

Their measure affected an array of industries, including coal, nuclear, oil, and gas. The plan that Bush sent to Congress June 28, 2001, was far more comprehensive than the bill that emerged in the House. The President's plan included provisions to reauthorize the 1957 Price-Anderson Act that limits legal liability of nuclear power plants in case of accidents; establishes mandatory reduction targets for certain power plant emissions; and develops broad electricity restructuring legislation.[79] President Bush felt confident that the United States could proceed in an environmentally friendly way.

At that time, Bush's energy plan encouraged the construction of more power plants and transmission lines. The report was specific about regulatory actions that could be taken and incentives given so that new plants would be built. Bush also planned to call for increased use of *eminent domain*, in which government can require the sale of private property for a project like a railroad, stadium, or power line. The issue is whether or not we should have the same authority on electrical transmission lines as the Federal Energy Regulatory Commission (FERC) has for gas pipelines,[80] Cheney told CNN. Eminent domain has never been granted, so that's one of the issues discussed. (Officials said that the Vice President decided to recommend the change.) Cheney observed that California had not built any electric power plants in the last 10 years and was experiencing rolling blackouts, because they simply did not have enough electricity. "They've got a whole complex of problems that are caused by relying only on conservation and not doing anything about the supply side," Cheney said.[81]

In March 2007, DOE submitted a $23.6 billion spending bill to Congress, a mere $45 million increase over the fiscal year budget request. It was designed to reduce our dependence on fossil fuels and emphasize investment in alternative fuel technologies that were stressed in the President's 2006 State of the Union Address. The Office of Environment Management received a $358 million increase to further DOE's commitment to safe cleanup of our Cold War-era nuclear facilities. Over half was for Rocky Flats, Fernald, Columbus, and Ashtabula sites. The Office of Nuclear Energy's NP 2010 program was funded at $80.3 million, allowing the Department to accelerate the engineering scope associated with the final designs. In addition, $167.5 million was for the Advanced Fuel Cycle Initiative (AFCI),[82] ongoing R&D, and technology development. The Office of Civilian Radioactive Waste Management would spend $445.7million in FY '07 and would perform the critical path activities needed to produce a high quality License Application for submittal to the NRC no later than June, 2008, that included completing certification of the License Support Network and the draft Yucca Mountain (YM) supplemental environmental impact statement.[83]

"The truth is that energy production and environmental protection are not competing priorities. They are dual aspects of a single purpose: to live well and wisely upon the Earth."[84] Nuclear power can eliminate the *forever-lasting waste and atmospheric emissions of fossil-fueled plants.*

It is important to note that commercial reactor operators have made great strides since 1976 in protecting their personnel from radiation, and the French, United States, and Navy experiences bear this out. In France, the average annual exposure of monitored French workers is 0.081 REM, while exposure from natural sources is much higher. And, U.S. Submariners and Aircraft Carrier personnel literally sleep with their reactors and do not suffer any ill effects. Radiation is not of major concern in modern reactors. The EPA continually monitors the environment: *Air* — particulates, ozone, carbon monoxide, and lead; *Land* — surface water, grass, and milk; and *Sea* — coastal water, algae, mollusks, and fish.

Concern for the environment led the French to the closed fuel cycle in lieu of long-term storage. France also developed and perfected all required technologies to perform the associated cleanup — its model is worth emanating. Unless the United States adopts a policy that utilizes spent fuel, it will continue to face billions of dollars in cleanup costs. The legacy of nuclear waste is the result of the national laboratories' "dole" or "entitlement" system and the insistence of the laboratories in "reinventing" existing technologies. Their mentality is *if it wasn't developed here, it is no good and must be redeveloped here.*

The environmentalists believe impacts of the warming of the atmosphere are becoming more evident and a greater problem worldwide. Carbon dioxide is one of several gases that contribute to trapping the infrared radiations emitted by the earth — radiations that cause global warming. Nuclear power generates neither carbon dioxide nor any other harmful or polluting gas. The environmentalists' position is that a massive reduction in carbon dioxide (CO_2) emissions throughout the world would mitigate climate changes and prevent rising sea levels. Today, nuclear power prevents 2.3 billion tons of CO_2 from being discharged globally every year. But, as previously indicated, carbon dioxide content in the atmosphere is increasing at a yearly rate of 0.5%, because of the consumption of fossil fuels and deforestation. This clear cutting of rainforests to create farmland and expand urban development, most notably in South America, is of major concern to environmentalist groups and others who worry about the future. If this pace continues, CO_2 content in the atmosphere will double in this century. Computer models estimate an average global warming of 1.5 to 4.5 degrees Centigrade (2.7 to 8.1 degrees Fahrenheit), causing dire consequences to the climate. The calculated relationship between this temperature rise and the CO_2 content in the atmosphere is referred to as the "Greenhouse Effect."[85]

Although concerned about global warming, environmentalists continue to oppose nuclear power. For example, Greenpeace, a non-profit organization with presences in 40 countries around the world, focuses on crucial worldwide threats to the planet's biodiversity and environment. A staunch critic of nuclear power, the organization has steadfastly maintained that nuclear power is the most complicated and dangerous means of producing electricity.[86] According to Greenpeace,[87] nuclear reactors are unsafe, uneconomical, and unnecessary. But, a profound transformation occurred on July 16th (2008) when Patrick Moore, co-founder of Greenpeace, who had led the anti-nuclear movement for decades, publically stated, "Thirty years on, my views have changed, and the rest of the environmental movement needs to update its views, too, because nuclear energy may just be the energy source that can save our planet from another possible disaster: catastrophic climate change."[88]

In 2004, France shut down its last coal mine and now gets 80% of its electricity from nuclear power. Because of this, France has the cleanest air of any industrialized country, and the cheapest electricity bill in all of Europe. Using spent fuel in the production of nuclear generated electricity has enabled *Electricite de France* (EDF) to reduce its CO2 emissions by 40%. According to the French Ministry for Ecology and Sustainable Development, atmospheric pollution has been reduced by 40%.[89]

Modeling after France, U.S. nuclear power can follow the logic of economic diversification and competitiveness. The advantages of nuclear power lie in generating clean energy by reprocessing and recycling used raw materials, concentrating its wastes, and providing an effective reduction in atmospheric pollution and global warming. Further, it will present no detectable adverse impacts on human health and wellbeing.

In the United States, recycling is accepted as an integral part of our lives, so why haven't we adopted this philosophy in regard to reactor fuel? A turnaround in our nuclear policy is being discussed, but a higher priority and a more responsible approach towards the environment must be adopted. Natural resources will not last forever — the "throw away mentality" is no longer acceptable. Sustainable development demands the conservation of an ecological balance by avoiding the depletion of our natural resources. France's reprocessing and recycling services are consistent with such policies, and they offer the United States' power generators a way of dealing with their spent fuel that is environmentally acceptable, saves natural uranium resources, and is commercially attractive. The United States should adopt France's recycling technology in dealing with the spent fuel issue.

CHAPTER 3
Situation

There are in fact two things, science and opinion;
the former begets knowledge, the later ignorance.

Hippocrates[90]

The problem of long-term disposal of nuclear waste, which is spent nuclear fuel from reactors and weapons-grade uranium and plutonium from dismantled weapons in the United States and Russia that is accumulating in large quantities in temporary storage sites, is both pervasive and politically charged. *The problem must be solved.* The issue of its disposition, whether to prepare the material for long-term storage or recycle it to use as fuel in mixed oxide (MOX) modified and a new generation of advanced reactors, is complicated. Expert opinion is divided. *The issue must be resolved.* The crux of the argument is whether the spent fuel and excess plutonium is a problem or an asset. The thrust of arguments presented herein is that nuclear waste is an asset, that it should be burned not buried, and that the United States should retake the lead in nuclear technology that it relinquished to France under the misguided policy of President Jimmy Carter. To reuse excess nuclear materials by recycling, spent fuel must be reprocessed to fabricate MOX. Weapons-grade plutonium, on the other hand, can be burned in the new generation advanced reactors — reprocessing is not required.

In July of 2008, the price of oil passed the $147 a barrel mark[91] (up from $20/barrel and continuing to rise), and the costs associated with reducing toxic emissions from coal-fired plants are so high that the cost-effectiveness of nuclear as an energy source is becoming increasingly attractive. A huge benefit, not easily reduced to a dollar value, is that reducing the nuclear waste stockpiles reduces the possibility that it fall into the hands of extremists. The government should support the nuclear industry in developing and implementing a long-term plan that increases the role of nuclear energy in order to reduce its dependence on foreign oil and dirty coal. In view of the continuing threat posed by extremists, consideration should also be given to construction of small, safe, and efficient *underground* nuclear reactors. In this plan, recycling must be embraced. In doing so, the United States could accomplish the following:

- *Influence public opinion regarding nuclear power plants;*
- *Effect new energy policies supportive of nuclear generation of electricity;*
- Reduce its dependence on foreign oil and, collaterally, the need for oil exploration in environmentally sensitive areas of the United States and its contiguous sea areas;
- Reduce environmental pollution by helping to eliminate green house gases, acid rain, and particulates, from oil- and coal-fired plants. (Millions of metric tons per year from coal plants, and the numerous deaths of coal miners.)

- Utilize recycled nuclear materials (U and Pu) from weapons being deactivated from nuclear arsenals of the United States and Russia for the generation of electricity; and
- Utilize reprocessed spent nuclear fuel in the generation of electricity as an alternative to long-term storage;

The accomplishment of these goals can be expedited by the following:

1. Focusing on recycling nuclear waste and nuclear reactor technology, and
2. Utilizing demonstrated expertise and experience of French nuclear utility companies and U.S. Navy nuclear propulsion program.

Books abound that address the development of highly industrialized nations, management, economics, nuclear energy, and electrical power generation and distribution. However, useful data on state-of-the-art reactors, hybrid nuclear fuels, recycling, and vitrification[92], are not well known except in the relatively small, tightly knit nuclear community. Subsequent to the terrorist attacks against the United States on 9-11, significant segments of information regarding nuclear technology, weapons systems architecture, and national military capabilities that routinely had been posted on the Internet were removed. There are limiting factors in the investigation of military capabilities and weapons systems, which are blocked, and *should be*, by security classification and access restrictions.

Also, information about the U.S. Navy's nuclear programs has been sparse; however, newspapers provide the most current information regarding ongoing debates on environmental issues, fears of nuclear attack by terrorist groups or some rogue-led[93] state, and energy policy issues. And, the Internet is a bountiful resource for general information and peripherals in the areas of environmental issues and political debates of the moment.

Fear that Islamic terrorists are planning an attack on U.S. nuclear power plants or DOE's nuclear facilities, based on current intelligence estimates, are of utmost concern. Officials familiar with these reports said that they contain potential methods of attack. Bombing or crashing an aircraft into a nuclear power plant or other nuclear facility, such as a weapons storage depot, could cause mass casualties and spread deadly radiological debris over a wide area.

President Bush said in his first State of the Union address that U.S. intelligence agencies found plans of U.S. nuclear power plants at terrorist bases in Afghanistan. They also discovered diagrams of American nuclear power plants and public water facilities, detailed instructions for making chemical weapons, surveillance maps of U.S. cities, and specific descriptions of landmarks. As a result, precautionary measures were taken and remain in effect to this day.[94]

A report released by the Department of Energy's (DOE) inspector general, stressed the point that there was "a weakness in controls over potentially dangerous materials" in government record keeping about plutonium and uranium that was loaned to U.S. academic institutions, private companies, hospitals, and other government agencies.[95] In many cases, the department could not account for nuclear materials loaned or leased to these licensees. For example: 1) substantial amounts of materials were located at two facilities that no longer existed; 2) several facilities carried negative balances that were not logical and almost certainly incorrect; and 3) nuclear records were incomplete in that they did not contain information on all government material provided to licensees. In the case of 119 locations, the management records showed licensees returned to DOE more nuclear materials than originally loaned or leased. In many cases, DOE believed the original transfer of the material was incorrectly reported. In 35 instances, there were more than 2,500 grams of plutonium reported returned.[96]

India and Pakistan have been engaged in a nuclear-charged face-off over Kashmir, part of it administered by India, Pakistan, and China. Although India seeks nuclear self-sufficiency, its ballistic missile programs are largely dependent on Russian components and expertise. In 1996, India suspended the Agni program under pressure from the United States, only to revive it in response to Pakistan's test of the Hatf-3 missile in1997. Both India and Pakistan conducted nuclear tests in May 1998, and both countries are actively pursuing their nuclear programs. At present, a version of the Agni is under development with a range up to 5,000 kilometers. Currently, India has only one type of nuclear-capable missile in service, the short range Prithvi, which can travel 90 to 150 miles and carry a payload of 1,200 to 2,400 pounds, the weight of a modest-size nuclear warhead. India has deployed these Prithvi missiles along its border with Pakistan.

China has the bomb. Its nuclear arsenal is in the midst of a speedy modernization program that began in the mid-1980s. By increasing the size, accuracy, range, and survivability of the nuclear arsenal, its leaders aim to strengthen its deterrent, duplicating the United States and Russia in deploying its nuclear weapons in a sea, air, and land triad. Our intelligence and defense agencies predict that within the next 15 years China will increase the number of warheads aimed at the U.S. targets from 20 to 75 or 100.[97]

North Korea and Iran want credible nuclear capabilities. In 2005 North Korea declared for the first time it had nuclear weapons. Shortly thereafter, in 2006, it conducted its first nuclear test with a Pu- based bomb. Iran, on the other hand wants nuclear status to establish regional hegemony and to deter further United States and Western influence.[98]

The world is a much more dangerous place today than it was 20 years ago during the Cold War. The new ingredient is *uncertainty*. No one can predict what may happen next, where, or when. In the spring of 1946, J. Robert Oppenheimer, director of Los Alamos National Laboratory, was asked in a closed Congressional hearing room "whether three or four men couldn't smuggle units of an (atomic) bomb into New York and blow up the whole city." The father of the atomic bomb answered, "Of course it could be done, and people could destroy New York." Then the nervous senator then asked how such a weapon smuggled into a crate or even a suitcase could be detected, Oppenheimer responded, "With a screwdriver."[99] A few years later, he persuaded the Atomic Energy Commission (AEC) to write a top-secret study on the dangers of nuclear terrorism. The report is known as the "Screwdriver Report," and it remains classified today.[100] Our leaders realized then that there was no defense against such an attack and that we are defenseless.

Then, September 11 happened and proved the point.

Oppenheimer understood a half century ago that by unlocking the power of the atom he and his colleagues had suddenly made the world a much smaller place and that is why he proposed banning all nuclear weapons. With globalization of science and technology, it has now reached a point where even a few individuals can wield nuclear weapons.[101] Even more frightening, perhaps, is the possibility that terrorists may attack the unsuspecting with a "dirty" bomb. As stated previously, current intelligence suggests that cells are here. They have the material and know how to use it. The question is not whether, but when and where.

It was reported on January 8, 2002, in the Washington Post, that two containers holding radioactive cesium-137 were stolen from a Russian industrial facility.[102] Four Georgians were apprehended trying to smuggle into Turkey two kilograms of uranium missing from the Russian city of Orenburg, a known black market source of plutonium. Nuclear materials are hot commodities in more ways than one. And now one can read about these unnerving episodes and many more in the mega-database of nuclear trafficking incidents posted on the Nuclear Threat Initiative's (NTI) web site.[103] Former senator Sam Nunn and media mogul Ted Turner's organization, the NTI, hopes to limit the spread of weapons of mass destruction through research, advocacy, public opinion, and grants. The NTI also works with the Henry L. Stimson Center and the Center for Strategic and International Studies.

The United States may be the last of the Super-Powers, but it's not the last to possess super-powerful weapons, powerful enough to wreak havoc on unsuspecting innocents.

At the peak of the Cold War, the United States and the Soviet Union each had at least 10,000 nuclear weapons.[104] Putin had initially signaled his intent to cut the number of Russian warheads down to 1,500 or less. This announcement in June 2001 in Ljubljana, Slovenia, was the result of a three-day summit between the two presidents. In a joint statement, they announced "a new relationship . . . founded on a commitment to the values of democracy, the free market, and the rule of law."[105] The United States and Russia had overcome the legacy of the Cold War. Neither country regarded the other as an enemy or threat. In recent years, however, relationships have become increasingly testy. Emerging from the post-Soviet economic meltdown, Russia resumed long-range strategic bomber flights, ending 15-years suspension of the missions.[106] Then, on August 8, 2008, the Russian Army marched into Georgia, and it seemed like déjà vu. Not to dwell on the subject, but the consequences of the Russian incursion could prove to be very serious over the long term.

At the close of the summit, nuclear differences remained between President Bush and President Putin. President Bush said, "We have a difference of opinion, but nevertheless, our disagreements will not divide us, as nations that need to combine to make the world more peaceful and more prosperous." Although both men vowed to trim nuclear arsenals by two-thirds earlier in the meeting, they did not agree on whether the weapons would be destroyed or whether the reductions would be permanent. However, Bush appeared to commit himself to destroying the nuclear warheads when he reduces the American arsenal. Arms control advocates criticized Bush's plan, because it did not specifically commit to the destruction of the decommissioned warheads. As part of their talks about nuclear proliferation, Putin and Bush discussed Osama bin Laden's efforts to obtain such weapons. Nuclear construction manuals found in Kabul (Afghanistan) depicted detailed designs for missiles, bombs, and nuclear weapons.[107] This discovery raised the specter of an attack that would far exceed the 9-11 atrocities.

During Presidents Bush and Putin's meeting in Shanghai, October 2001, while standing side by side at a news conference, Bush argued that the September 11 terrorist attacks on New York and Washington made a missile defense system more urgent. Their failure to agree was on the Anti Ballistic Missile (ABM) Treaty, which prohibits the United States and Russia from building nationwide defenses against ballistic missiles. Bush said that the missile defenses would protect both countries from nuclear blackmail and potential terrorist attacks. Shortly after that meeting, on June 2002, six months after giving the required notice of intent, the United States withdrew from the treaty. This led to the creation of the Missile Defense Agency that is responsible for developing a layered defense against ballistic missiles.[108]

President Bush has named Iran and North Korea as the *axis of evil.*[109] However, Russia and China sell the means to produce and deliver weapons of mass destruction. Both countries sell to Iran. A Pakistani engineer, A.Q. Kahn, is known to have sold nuclear knowledge and materials on the international black market, and this is a matter of concern because of the high probability that terrorists have access. China also sells equipment and technology for producing biological and chemical weapons. These are serious matters, particularly because Iran is well along in its nuclear program, as well as in the production of chemical and biological weapons. And, it has ballistic missiles to deliver them. These could be passed on to terrorists. North Korea is considered the world's number one merchant for ballistic missiles, open for business with anyone.[110]

Former Defense Secretary Donald H. Rumsfeld said that the United States could face terrorist action "vastly more deadly" than 9-11, particularly from terrorists with ballistic missiles armed with chemical, biological, or nuclear weapons that could kill thousands. Of interest is the joint DoD and DOE development of the Robust Nuclear Earth Penetrator, also known as the *bunker buster.* This low-yield, strategic nuclear weapon, if employed against stores of chemical and biological weapons, would destroy, not only the containers, but also the agents themselves. Conventional weapons, however, would destroy the containers and simply "free" or release the agents, which would be dispersed initially by the blast and subsequently by the wind.[111] The real nexus is between terrorist networks and terrorist states that have weapons of mass destruction.

"Let there be no doubt: There is that nexus."[112]

CHAPTER 4
The ABCs of Nuclear

The more a subject is understood,
the more briefly it may be explained.
Thomas Jefferson[113]

Uranium Fuel

The raw material used to make nuclear fuel is uranium ore. More plentiful in nature than once thought, it has few other commercial uses except certain special applications in aerospace industries. The countries that provide uranium listed in descending order of 2006 world production of 39,429 tons are as follows: Canada (25%), Australia (19.3%), Kazakhstan (13.4%), Niger (8.7%), Namibia (8%), United States (4%), China (1.9%), South Africa (1.4%). Others at less than one percent (<1%) include the Czech Republic, India, Brazil, Romania, Germany, Pakistan, and France.[114] The largest producing uranium mines in 2006 were Canada's McArthur River (underground) with 17% of the world total, Australia's Ranger (open pit) with 11%, and Australia's Olympia Dam (open pit) with 8%. Nearly half of the world's production of uranium comes from mines in Canada, Australia, and Kazakhstan.[115]

To date, natural or slightly enriched (3 to 5 %) uranium is used to fuel all commercial nuclear power reactors. In natural uranium ore, there is only one atom of the fissionable isotope present for every 140 atoms of non-fissionable isotope. Only the U235 atom can be split to produce energy, and less than one percent (0.7%) by weight of the uranium mined is fissionable and therefore useful. The rest (99.3%) of the uranium consists of a trace of the U234 isotope and the non-fissionable isotope U238 that has a half-life of 4.5 billion years. On site or proximate to the mines, natural uranium ore is milled and refined to extract *triuranium octoxide* (U_3O_8, usually called simply *uranium oxide*), which is purified, concentrated, and transformed to yellow powder, called *yellow cake.* Yellow cake typically contains 70 to 90% uranium oxide, and it is in this form that uranium is marketed.

Subsequently, yellow cake is transformed into gas, uranium hexaflouride (UF_6), by a chemical process, and then enriched to increase the concentration of U235 to 3 to 5% for nuclear fuel. The nuclear industry uses two methods of enrichment: gaseous diffusion and gas centrifuge. Both methods use the difference in molecular weight of the U235 and the U238 isotopes to alter the ratio of the two in the gaseous mix. Uranium is enriched by increasing the number of lighter more easily split isotope U235. (The gaseous diffusion process for enrichment is used at the Paducah (Kentucky) Gaseous Diffusion Plant. The only operating uranium enrichment facility in the United States, the plant is operated by the United States Enrichment Corporation (USEC) Inc., a global energy company. However, the gas centrifuge process has been used for more than 30 years in Europe to enrich uranium for the commercial nuclear power market. In June 2006, the U.S. Nuclear Regulatory Commission (NRC) issued a license to Louisiana Energy

Services (LES) to construct a commercial gas centrifuge facility in New Mexico.[116])

The enriched uranium is then transformed to uranium oxide (U_3O_8) in the form of brown powder, which is compacted into small pellets that weigh only 7 grams each but contain an enormous amount of energy. Two pellets, about 15 grams, possess the energy equivalent of one *tonne* (2,205 pounds or about eight barrels) of oil! The pellets are stacked in tubes, called *fuel tubes*, and grouped into bundles that are called *fuel assemblies*. Fuel assemblies undergo a nuclear chain reaction and, thereby, supply energy in the form of heat. The U_{235} is gradually consumed by the fission reaction, and usually one third of the fuel assemblies are replaced every three to four years. This reloading operation can be undertaken only when the reactor is off line, or shutdown; however, it can be done on line with Canada Deuterium Uranium Reactor (CANDU) or Pebble Bed Modular Reactor (PBMR) reactors.

The spent fuel assemblies that are 'hot' in both thermal heat and radioactivity are placed in pools to cool down. In addition to the thermal cooling provided, water absorbs radiation in the sense that it is a barrier that stops the propagation of radioactivity. This nuclear waste can be considered a liability or an asset — the United States considers spent fuel a liability; others, most notably France, but also Germany, Russia, Belgium, England, and Japan, consider spent fuel an asset. *As a liability*, it must be isolated permanently in a storage facility, such as Yucca Mountain, but a*s an asset*, it can be reused over and over again, providing energy and reducing the final quantity of waste destined for long-term storage to a fraction, perhaps 10%, of what it would otherwise have been. Research on new approaches to full recycle or closed fuel cycle is ongoing in France, Japan, Russia, and the United States. Advanced technologies research at the Argonne and

Idaho National Laboratories could lead to processes that result in waste forms containing less than 1% transuranics and no fission products.[117]

To reprocess spent fuel, plutonium is recovered and mixed with uranium. The new mixed oxide fuel, called MOX, is now ready to be burned in present light water reactors (LWRs) or the new-generation breeder reactor. (The MOX fuel is presently used in European reactors, although not in a full core load.) Approximately 20% MOX is inserted with 80% uranium fuel. Eventually, full core loads of MOX will be utilized. Duke Energy's Catawba Nuclear Station, South Carolina, has already completed irradiation of MOX using weapons-derived plutonium lead test assemblies (LTAs) that are basically fuel bundles comprised of MOX fuel rods.

Nuclear Fission

Enrico Fermi of Italy irradiated uranium with neutrons in 1934. He believed he had produced the first transuranics element, but in reality he had discovered fission. In 1938 two Germans, Otto Hahn and Fritz Strassman split the uranium atom with neutrons and showed that barium and krypton elements were formed by the process. Unaware of it at the time, they had actually induced a fission reaction. Later research by Lise Meitner (a colleague of Hahn) and her nephew Otto Frisch,[118] established the knowledge that would lead to its applications, first in weaponry and then in peaceful initiatives, including nuclear power.[119] Meitner recognized the possibility for a chain reaction of enormous explosive potential. Because this had the potential to be used as a weapon and the knowledge was in German hands, Edward Teller, Leó Szilárd, and Eugene Wigner together persuaded Albert Einstein to write a letter alerting President Franklin D. Roosevelt. This warning letter led directly to initiation of the Manhattan Project.[120] (The development of the first nuclear weapons by the Manhattan Project led to the surrender of Japan that ended World War II.) Application of nuclear energy in the commercial power industry followed shortly after the war.

The 92 chemical elements are listed by the number of protons in their nuclei, from hydrogen (H) that has one to uranium that has 92. Each atom of an element has an equal number of protons (positively charged) and electrons (negatively charged), but the electrons are so tiny as to be counted zero weight. The mass number of the element is the sum of its protons and neutrons. Uranium is one of the few materials capable of producing heat in a self-sustaining chain reaction.

The fission process starts with a neutron striking the target nucleus of an atom whereby the neutron is absorbed. This absorption increases the energy level of the target nucleus and "splits the atom." The resulting products are more neutrons to fission other targets, fission products (different radioactive elements), and heat. When formed, the fission products initially move apart at very high speeds, but they travel only a few thousandths of an inch before they are stopped within the fuel cladding. Most of the heat produced in fissioning comes from stopping these fission products within the fuel and converting their kinetic energy into heat.

In the case of uranium, when a neutron strikes a U_{235} nucleus, uranium *fissions*. The nucleus splits producing two fission fragments of lighter elements and two or three new neutrons. Each fragment consists of a nucleus having approximately half the neutrons and protons of the original nucleus. The most abundant (0.71%) *fissionable* isotope of uranium is $U_{235,}$ but the most abundant (99.28%) isotope U_{238} is *non-fissionable*.

The *reactor core* is where the nuclear fissioning takes place. Its four components are the fuel assemblies, control rods, moderator, and coolant. (Outside the core of the reactor are the heat exchanger, turbine(s), and the external component of the cooling system.) The fuel assemblies are composed of a large number of fuel rods each of which contains hundreds of stacked uranium fuel pellets. The uranium atoms are sealed within the fuel rods by metal cladding that is highly resistant to erosion and radiation. The control rods also contain pellets of neutron *absorbers,* such as boron (B) or cadmium (Cd) to control the reaction rate by being inserted to slow or stop fission or withdrawn from the core to increase fission and start a chain reaction. The moderator, usually water (light or heavy) or helium gas, slows the high speed neutrons that are bouncing about in the core. The coolant is usually water, and in some reactors the moderator and coolant are one and the same.

When an uranium-235 (U_{235}) atom fissions, neutrons are produced in one ten-thousandth of a second. These *fast neutrons* tend to be captured without fissioning by the uranium-238 (U_{238}).[121] It is only the slower *thermal* neutrons that cause the U_{235} to fission. The moderator slows the speeding neutrons so they can bounce off without getting caught or escaping, and the fission can begin. The movement of a control rod changes the power level. Increases in reactor power can be obtained simply by pulling out a sufficient amount of control rods. Precise control is needed to produce a steady reaction at the desired rate.

By slowing down fast neutrons through collisions with light elements, such as water, they react more readily with U_{235} atoms. This property is used in "thermal neutron" or slow reactors. It reduces the enrichment in U_{235} required for the chain reaction. In water reactors, the slowing material or moderator water is also the fluid transporting the heat or coolant.

Each neutron expelled upon fission (splitting) of a U_{235} atom may collide with another atom, causing its own fission, accompanied by a new release of energy and a neutron, which causes another and another. This is a chain reaction with resulting heat typically used for power generation.[122] Because it reacts with neutrons, U_{235} can sustain a chain reaction even when in small proportions. This reaction is transmitted at a very high speed from one atom to another, giving considerable added energy. *The fission reaction of one pound of U_{235} can supply as much energy as burning 6,000 barrels of oil.*[123] These two phenomena — nuclear fission and chain reaction — are used in a nuclear power plant reactor. The energy released by fissioning this fuel is recovered as heat and turned into electricity through a water vapor or gas cycle.[124]

One anomaly of neutron physics is that a small percentage of neutrons is not released instantly with the fission reaction. These "delayed" neutrons, which make the spent fuel thermally hot, have to be included in the fission calculation. If the neutron population can be controlled, a steady chain reaction can be maintained, or it can be shut down immediately.

Running, controlling, and cooling the fission in the reactor core are accomplished through three components: the moderator, control rods, and coolant. Various combinations of these three determine the types of reactors or the system. Several combinations have been tested, but only a few have passed the stage of "prototype facility" to achieve industrial electric power production and operation. Two reactors, Boiling Water Reactor (BWR) and Pressurized Water Reactor (PWR), are very similar in concept and can be used to explain very simply the process whereby nuclear energy is harnessed to produce electricity. Each system is composed of a nuclear core, control rods, turbine, generator, and condenser. In both cases, the generator is connected to the electrical grid that ultimately delivers electricity to the user.

Safety

Like any conventional thermal or heat plant, a light water reactor (LWR) plant has a "boiler" (the reactor) that transforms water into steam. The steam drives the turbine, which drives the generator that produces electricity. After steam passes through the turbine, it is cooled and returned to the reactor. The cycle continues. Because radioactive materials can be dangerous, nuclear power plants have safety systems that protect the employees, the public, and the environment. If necessary, the reactor can be shut down quickly to stop the fission process; the reactor core can be cooled down and heat carried away from it; and the radioactivity can be contained to prevent it from escaping or "leaking" to the environment.

The only section of a nuclear power plant where significant radioactivity is present is in the reactor itself. Under normal reactor operation, neutrons are present *only* in the reactor core. Radiation is constrained in the fuel rods by cladding, the LWR coolant (water) itself is a barrier, and the 9–inch-thick reactor pressure vessel is designed to contain radioactivity. Most of the neutron flux is contained within the pressure vessel. Components that are of most concern are the nuclear fuel rods, the pressure vessel, and the containment building.

Protective Barriers

The uranium fuel *pellets* are stacked in welded metal rods, the walls of which have cladding. These fuel rods are grouped by the hundreds into *bundles* called *fuel assemblies*. The fuel assemblies are placed into a reactor *pressure vessel* that has walls made of thick steel plates that are welded together. The reactor pressure vessel is located in a leak proof *containment* building that has walls made of reinforced concrete or steel, or a combination of both.

In the sequence of the process, the *first* protective barrier is the *reactor fuel cladding* or the walls of the fuel rod. The *second* (sometimes called primary) barrier is the walls of the *reactor pressure vessel.* The *third* (sometimes called secondary) barrier is the walls of the *containment building* itself. Nuclear power plants vary in design, but *safety always is a principal design criterion.*

Plutonium

Approximately 95 to 97% of the fuel in the reactor is U_{238}. Some of this uranium transforms into plutonium-239 and plutonium-241. As the plutonium aggregates over time, the uranium decreases to a level at which the reactor requires refueling.[125] *All plutonium is man-made.* It is formed in the reactor when U_{238} captures a neutron and transforms into U_{239}. In turn, U_{239} transmutes into neptunium-239 (Np_{239}). Then, every other day, half of the Np_{239} transforms into Pu_{239}. (Pu_{239} decays very, very slowly, having a half-life of 24,110 years.) About 30% of the power produced by the reactor comes from the plutonium. At the end of the cycle, approximately two years, some U_{235} is left.

To some people, plutonium is viewed negatively because of its potential use in weapons of mass destruction, *a threat.* The often-used false statement, "one atom of plutonium can kill you" has given the word plutonium another ominous definition). But to others, plutonium is viewed positively because it is a source of energy, *a resource.* Based on projections of fuel assemblies to be discharged as spent fuel through the year 2030, *there will be enough plutonium in the U.S. spent fuel assemblies to operate 20 reactors for 40 years.*

Thorium Fuel

Thorium (Th [atomic number 90]) is a metal actinide slightly lighter than uranium (U [atomic number 92]). Much more abundant than uranium, large reserve deposits are found in Australia (25% [of

71

world total]), India (24%), Norway (14%), Canada (13%), and the United States (8%). The two elements, uranium and thorium, have many of the same characteristics; most importantly, both can absorb neutrons and transmute into fissile elements. Natural thorium (Th_{232}) absorbs a neutron and quickly transmutes into unstable Th_{233} and then into protactinium (Pa_{233}) before decaying into U_{233}. The result of this complicated process is that the U_{233} that is produced at the end of this breeding process is similar to U_{235}. Thorium-232 is *fertile*, therefore it is practical as a nuclear fuel. But one significant difference between the two elements, and one that has brought renewed interest, is that thorium is not per se *fissile*. It cannot *fission* on its own, no matter how tightly packed. Because it cannot sustain a nuclear chain reaction, thorium cannot be used to make a nuclear bomb!

An advantage over uranium fuel is that thorium does not produce as many highly radioactive byproducts, and no plutonium is bred in the fission process. Of extraordinary consequence is that the thorium fuel incinerates plutonium and other actinides thereby reducing the stockpiles of spent fuel and excess weapons-grade plutonium held in temporary storage.[126] Additionally, the high-level waste produced using thorium fuel is a fraction of that produced by uranium-fueled reactors, and the waste remains radioactive for less than 500 years rather than 10,000 years. The long-term storage problem is not solved, but is reduced dramatically. Thorium fuel technology was validated by successful operation at the Shippingport Light Water Breeder Reactor (LWBR) in the early 1980s.

Thorium Power, Ltd., McLean, Virginia, a nuclear consulting firm, is developing thorium-fueled nuclear power. Its approach to the sub-criticality of thorium is to mix it with enriched uranium and plutonium. The center of the fuel rod is plutonium, which serves as the "seed" that generates the neutrons necessary to start the thorium fuel cycle. Wrapped around the core is a "blanket" containing a mixture of uranium and thorium that also fissions. Thorium Power has run successful tests of its fuel in a research reactor in Moscow.

If thorium fuel is accepted and put to use commercially in the United States, it will accomplish two goals: dispose of weapons-grade plutonium (without producing plutonium in the process) and generate electricity.

Light Water Reactors (LWRs)

Light water reactors are also called *water moderated reactors* because the water serves as a moderator as well as a coolant. As in any conventional thermal or heat plant, a nuclear power plant consists of a boiler, which transforms water into steam. The steam drives a turbine, which in turn drives a generator to produce electricity. After steam passes through the turbine, it is returned to the reactor. In nuclear power plants, the only section of the plant where significant radioactivity is present is in the reactor itself. Components of the reactor that are of concern are the nuclear fuel rods, the cooling system, and the containment vessel. (Under normal operation, neutrons are present only in the reactor core.) Radiation is constrained in the fuel rods by cladding, the coolant itself is a barrier, and the 9 – inch thick reactor pressure vessel is designed to contain radioactivity. Most of the neutron flux is contained within the pressure vessel. Because the building housing the reactor is built to strict standards of constructional integrity, there are in fact four barriers to radioactivity: nuclear fuel cladding, coolant, containment vessel, and the containment building.

The fuel rods and pellets in the reactor core have cladding, which is contained in a 9-inch-thick pressure vessel. This pressure vessel, in turn, is contained in a solid, leak-proof containment building, meeting the stringent requirements of nuclear safety. Thus, the fission products are contained within four different barriers (pellets, rods, pressure vessel, and containment building).

In the Boiling Water Reactor (BWR) heat generated by fissioning of nuclear fuel in the core literally boils the water. The steam generated drives the turbine, which in turn, spins or drives the shaft of the generator. The electricity thus produced is distributed through the electrical grid. Internally, having powered the turbine, the steam is condensed and the water is used again as the cycle continues.

The Pressurized Water Reactor (PWR), on the other hand, has three separate water "loops" in independent piping systems. In the first loop, water is heated by nuclear fuel, but it is pressurized so that it does not boil. The pressurized superheated water is run through a steam generator where un-pressurized water in the second loop is heated to the boiling point and converted to steam that drives the turbine-generator to produce electricity. The steam then passes to a condenser, where cooling water in the third loop converts the steam to water, and the process continues. The advantage of the PWR is that the water used to generate steam never contacts the pressurized water heated in the reactor core, so there is no opportunity for external contamination.

Nuclear power plants must have a *cold sink* to remove the heat generated by the fission process. Because water often is the coolant of choice, nuclear power plants usually are built on the seashore, an island, a river, or a lake. Many nuclear and coal power plants are equipped with cooling towers where droplets of water evaporate, transferring the heat to the atmosphere rather than to a body of water. Condensation coming from nuclear power plant cooling towers is not radioactive.

Classification of Nuclear Reactors

Classification by *type of nuclear reaction*
- Nuclear fission. All commercial reactors are based on nuclear fission.
    - Thermal reactors use *slow* neutrons, and most power reactors are of this type.
    - Fast neutron reactors use *fast* neutrons to sustain fission.
- Nuclear fusion. Fusion power remains an experimental technology.

Classification by *moderator material* (used by thermal reactors):
- Water moderated reactors
    - Light water reactor (LWR)
    - Heavy water reactor (HWR)
- Light element moderated reactors
    - Molten salt reactor (MSR)
    - Liquid metal cooled reactors

Classification by *coolant*
- Water cooled reactors
    - Boiling water reactor (BWR)
    - Pressurized water reactor (PWR)
        - Most commercial PWRs
        - Naval reactors
- Liquid metal cooled reactors
    - Sodium-cooled fast reactor
    - Lead-cooled fast reactor

- Gas cooled reactors
  - Helium
  - Nitrogen
  - Carbon dioxide
- Molten Salt Reactors (MSRs)

Classification by *use*
- Electricity
  - Nuclear power plants
- Propulsion
  - Nuclear marine propulsion
- Other uses of heat
  - Desalination
  - Heat for domestic and industrial heating
  - Hydrogen production for use in a hydrogen economy
  - Production reactors for transmutation of elements
  - Breeder reactors.
  - Production of weapons-grade plutonium
  - Creation of radioactive isotopes
- Research reactor: Typically very small reactors

Classification by *generation*
- Generation I reactor
- Generation II reactor
- Generation III reactor
- Generation IV reactor

Classification by generation will be used in further discussion of nuclear reactors because it gives a sense of the history of reactor technology development and application.

Nuclear Fuel Cycle

Before and after its use in the core of reactors, nuclear fuel undergoes operations and transformations making up the nuclear fuel cycle. The *front-end* of the cycle involves the following:

- Prospecting for uranium deposits
- Mining
- Uranium ore processing (milling)
- Chemical conversion of uranium to yellow cake
- Enrichment in U_{235}
- Manufacturing of fuels

In a closed fuel cycle, the *back-end* involves:

- Reprocessing of spent fuels
- Recycling of reusable materials
- Conditioning and interim storage of waste

The fissile materials, uranium and plutonium, recovered in recycling can be used in the manufacturing of fuels, either uranium-based fuels or MOX fuels. In an open fuel cycle, the spent fuel is simply stored for eventual disposal in a geologic repository.

Nuclear energy is mankind's only non-greenhouse-gas emitting electrical power source, available regardless of weather or time of day. It is adept to producing electricity 24/7, and it is mankind's cleanest, most-efficient, and reliable source of electricity. Of increasing interest is the use of nuclear reactors in the production of hydrogen as a source of energy in the 21st century.

CHAPTER 5
The United States Navy

In all of history "he who commands the sea,
has command of everything."
Cicero[127]

For decades after World War II, the United States faced an uneasy peace during which time the Navy was coping with a seemingly continuum of international crises. In the early 1950s, the Navy had only 634 active ships. Crews spent too much time at sea and too little time on the upkeep of their ships. The pressures of exercises increased and, because of rapidly changing technology, so did the demands of proficiency training. The long wars in Korea and Vietnam were costly, politically limited, and socially divisive. The Navy suffered.

World War II had ended with the deployment of the nuclear bomb — it is obvious that will be its all-time legacy. No one thought about the implications that nuclear power might have *for* submarines, instead of *against* them, except a slight-of-build "Engineering Duty Only" captain by the name of Hyman George Rickover. Assigned to the Navy Bureau of Ships in Washington, DC, Captain Rickover was given authority to look into nuclear power for propulsion of ships. Thanks to him, today's United States Navy nuclear-powered ships and submarines are the most powerful in the world.[128]

Rickover's foresight exceeded that of all his contemporaries as to what a nuclear power plant could do for the submarine and the Navy's major surface warships, the aircraft carriers. The best World War II-era submarines, with postwar improvements, were capable of 20 knots on the surface, and they could range about 10,000 miles on a full load of diesel fuel. Submerged, they could make 15 knots for about an hour, after which the boat would be flat with the battery completely discharged. At very slow speed, a submerged submarine hunted by enemy anti-submarine units could stay down for about 48 hours. By contrast, a submarine with a nuclear power plant would be capable of 30 or more knots, whether surfaced or submerged, for several years.

In unsupported cruising range, Rickover brought the Navy back to the days of the sail. *A nuclear reactor is nearly as inexhaustible as the wind.*

"Underway on nuclear power"[129]

Those were the words uttered from the USS Nautilus (SSN 571) when she put to sea for the first time on January 17, 1955.[130] Made possible by the Navy, government, and contractor engineers led by Captain Rickover, this signal event revolutionized undersea warfare. Only nine years earlier, Rickover had obtained Congressional support to develop an industrial base in a new technology, pioneer new materials, design, build, and operate a prototype reactor, establish a training program, and take a nuclear-powered submarine to sea. This new industrial base in nuclear propulsion has given the United States Navy sea and air supremacy ever since.[131]

The Navy Bureau of Ships was responsible for the design of surface ships and submarines, but the Navy had to work with the Atomic Energy Commission (AEC), because it was responsible for reactor development. Rickover arranged for several officers, civilians, and himself to be transferred to Oak Ridge, Tennessee, to learn the basics or fundamentals of nuclear physics and technology. Rickover was on a mission. He returned to Washington and lobbied the Congress to establish a nuclear propulsion program for the Navy.[132]

The Naval Nuclear Propulsion Program (NNPP) was established in 1946, shortly after the conclusion of World War-II. Its headquarters provides oversight and direction for all elements of the program. Because of the critical nature of the program, the professionals at this office make all major technical decisions regarding design, procurement, operations, maintenance, training, and logistics. The NNPP sets the standard for excellence in the efficient, reliable, and safe operation of nuclear reactors. That same year, Congress passed the Atomic Energy Act (AEA) that created the Atomic Energy Commission (AEC) within the DoD. The AEC succeeded the wartime Manhattan Project and was given the responsibility for developing atomic energy.[133] Thus it came to be that the Naval Nuclear Propulsion Program fell under the purview of the Secretary of Defense, while the civilian nuclear initiative, the Naval Reactors Program, was later to fall under the purview of the Secretary of Energy.

The concept of using a reactor to produce heat was understood; however, the technology to build and operate a nuclear propulsion plant on ship did not exist. Several reactor concepts were on the drawing board, but the real challenge was to develop the technology and transform theory into practice. New materials had to be developed, components designed, and fabrication techniques had to come together. Installing and operating a steam propulsion plant inside the confines of a submarine and under the sub-sea pressure conditions were mind-boggling. The team at Oak Ridge, Tennessee, knew that to build a Naval nuclear propulsion plant would require a huge commitment of resources and a forward thinking government and industry.

The Navy created the Nuclear Power Branch on August 4, 1948. Rickover recruited a strong, committed, technical staff that studied at Oak Ridge, choosing civilian engineers and Navy officers to oversee every aspect of the development of nuclear propulsion. He recommended undertaking two parallel reactor development projects: a pressurized water reactor, described earlier, and a liquid metal cooled reactor. By 1949, Rickover had an arrangement with the AEC and the Navy to proceed with both reactor projects.[134]

The Navy contracted with the Westinghouse Electric Corporation to construct a new facility and develop a prototype submarine propulsion plant. The Bettis Atomic Power Laboratory was founded that year in Pittsburg, Pennsylvania, to construct, test, and operate a prototype submarine reactor plant utilizing the pressurized water design. This first reactor plant was called the Submarine Thermal Reactor (STR). Then, in 1950, Rickover contracted with General Electric to work on the liquid metal reactor design, which was already underway for the AEC at the Knolls Atomic Power Laboratory, in upstate New York.[135]

On March 30, 1953, the Submarine Thermal Reactor (STR) was brought to power for the first time, and the age of naval nuclear propulsion was born. One of the greatest revolutions in the history of naval warfare had begun. To test and operate this reactor plant, Rickover put together an organization that has lasted nearly 60 years. The Bettis Atomic Power Laboratory was assigned responsibility for operating the reactor it designed and built. Naval personnel trained the cadre crew. Admiral Rickover ran a tight organization, which ensured safe operation of the nuclear reactor plant, because he enforced the strictest standards of technical and procedural compliance.[136]

The successful development of the nuclear propulsion plant led to the construction of the Navy's first nuclear submarine, the USS Nautilus (SSN 571), which utilized the pressurized water design. The Nautilus' first sea trial in January of 1955 marked the transition of submarines from slow underwater vessels to fast 25- to 30-knot warships. The USS Seawolf, powered by the Liquid Metal Reactor (LMR), was commissioned in 1957. She operated at sea for two years, but the pressurized water-cooled design proved to be the preferred reactor for Navy's nuclear program.[137] An aircraft carrier, the USS Enterprise (CVN 65), powered by eight reactor units, was commissioned in 1961 and remains in service to this day. The USS Long Beach, a cruiser powered by two reactor units, followed in 1961. In 1962 the Navy had 26 nuclear submarines operational and 30 under construction.

Clearly, nuclear power had revolutionized the United States Navy.[138]

The government moved the Naval Nuclear Propulsion Program from the Atomic Energy Commission to the Department of Energy in 1970, but it retained its dual agency responsibility with Department of Defense. The NNPP's basic organization, responsibilities, and technical discipline have remained just as Rickover structured them more than two decades earlier.

The Naval Nuclear Propulsion Program's success is based on strong, central, technical leadership with thorough training, conservatism in design and operating practices, and a complete knowledge of every aspect of the program. (This sentence encapsulates the reasons that the U.S. Navy Nuclear Propulsion Program was chosen as the model for this book.) Rickover expressed it this way: "Responsibility is a unique concept: it can only reside and inhere in a single individual. You may share it with others, but your portion is not diminished. You may delegate it, but it is still with you. You may disclaim it, but you cannot divest yourself of it. Even if you do not recognize it or admit its presence, you cannot escape it. If responsibility is rightfully yours, no evasion, or ignorance, or passing the blame can shift the burden to someone else. Unless you point your finger at the person who is responsible when something goes wrong, then you have never had anyone really responsible." Rickover taught his crew well. He maintained that excellence must be the norm in the Naval Nuclear Propulsion Program. To maintain these standards, individuals must accept responsibility for their actions.[139]

The former governor of Idaho, Dirk Kempthorne, said, "…Since the days of Admiral Rickover, the men and women of the Navy Nuclear Propulsion Program have been recognized around the world for their high standards of achievement and performance, their commitment to professionalism, and their dedication to accountability."

Thorough training minimizes problems. This results in quick and efficient responses to emergencies that assure safety. Plant operators must meet tough selection standards and complete extensive nuclear propulsion training and qualifications successfully before reporting to a submarine or carrier. They must be fully qualified in areas of thermodynamics, reactor principles, radiological fundamentals, and other specialized areas. Training, which is hands-on, is a way of life in the Nuclear Navy.

The NNPP is made up of military personnel and civilians, who design, build, operate, maintain, and manage the nuclear-powered ships and the many facilities that support the U.S. Navy nuclear-powered fleet. The program has responsibility for nuclear propulsion from cradle to grave. The program includes:

- The Naval Nuclear Propulsion Program Headquarters organization and field offices;
- Research and development laboratories;
- Contractors responsible for the design, procurement, and construction of propulsion plant equipment;
- Nuclear power schools and naval reactor training facilities;
- Shipyards that construct, overhaul, and service the propulsion plants of nuclear-powered vessels;
- The Navy's nuclear-powered warships;
- Navy support facilities and tenders; and
- Storage of spent nuclear fuel and reactor plant disposal.

The government-owned, contractor-operated Bettis and Knolls Atomic Power Laboratories are research and engineering facilities dedicated to the Naval nuclear propulsion work. (Bettis is located in Pittsburg, Pennsylvania; Idaho Falls, Idaho; and, Charleston, South Carolina. Knolls is located in upstate New York.) They staff 5,400 personnel to develop the most advanced Navy nuclear propulsion technology and provide technical support for the continued safe and reliable operation of all existing Naval nuclear reactors. The Knolls Atomic Power Laboratory in New York tests the new designs and promising new technologies under typical operating conditions before introducing them into the Fleet.[140]

In the early 1950s at an 890 square-mile area in southeastern Idaho, research was initiated at the Idaho National Engineering Laboratory (INEL), now known as the Idaho National Laboratory (INL), to develop reactor prototypes for the U.S. Navy. The Naval Reactors Facility (NRF), a part of the Bettis Atomic Power Laboratory, was established to support the development of the Naval Nuclear Propulsion Program. Under the direct supervision of the DOE's Office of Naval Reactors, the Westinghouse Electric Corporation operates the

Naval Reactors Facility. The facility supports the NNPP by carrying out testing, examination, and spent fuel management activities. In 1954 the Navy's presence expanded to eastern Idaho where the Nuclear Power Training Unit was established.[141]

Although they have been deactivated, there are three naval nuclear reactor prototype plants at this facility. The Expended Core Facility (ECF) and support buildings were constructed in 1958 and are still in operation. They receive, inspect, and conduct research on naval nuclear fuel.[142] The submarine thermal reactor prototype was operational for 38 years; the large ship reactor prototype was operational for 36 years; and the submarine reactor plant prototype was operational for 30 years. Sailors were trained for the Navy as well as for research and development (R&D) purposes.

The Test Reactor Area (TRA), now called the Reactor Technology Complex (RTC), occupies 102 acres of the southwest portion of the INL. In the early 1950s, the TRA was constructed and designated as the Materials Test Reactor that was shut down in 1970. Two other reactors were built at the TRA: The Engineering Test Reactor, which was deactivated in 1982, and the Advanced Test Reactor that is still in operation and utilized extensively by DOE's Naval Reactors Program (NRP). After nearly 40 years of operation, the ATR is still considered a premier test reactor and is contributing to research, radiation testing, and isotope production.[143]

In the late 1950s, the Plant Apparatus Division (PAD) and the Machinery Apparatus Operation (MAO) were established to provide engineering, procurement, and technical oversight of Naval nuclear components. Many privately owned companies in the United States perform the actual design and fabrication of the major propulsion plant components. The manufacturing of the heavy components used in Naval reactors requires four to five years of precision machining, welding, grinding, heat treatment, and nondestructive testing of large, specialty metal forging under very carefully controlled conditions. The standards set forth by the Navy are far more rigorous and stringent than those required for civilian nuclear reactors. Components on warships must be designed and built to accommodate battle shock, crew proximity to the reactor, and frequent, often rapid, changes in reactor power, while preventing radiated noise.

The INL receives spent nuclear fuel (SNF) from DOE reactors in other states, U.S.-supported reactors in other countries, commercial nuclear power plants, and reactors from U.S. nuclear-powered submarines and aircraft carriers. The Navy ships spent fuel from the shipyards that support its nuclear ships and boats to the INL. This process is necessary to meet national security requirements to defuel Navy reactors, and it ensures examination of fuel from these sources.

A Federal Court settlement agreement in a case of the United States v. Batt, the Governor of Idaho, in October 1995 certified the total number of shipments of naval spent fuel required through the year 2035. The Navy will not ship more than 24 shipments to INL from the date of this Agreement through the end of 1995, 36 shipments in 1996, 20 shipments each year from then until 2035. The total number shall not exceed 575 and shall not exceed 55 metric tons of spent fuel. As a result of these constraints, spent fuel inventories continue to build at the yards.[144]

Admiral Rickover headed the NNPP for 35 years until his retirement in 1982. He left a legacy that has been continued by his successors. The result is a fleet of nuclear-powered warships unparalleled in capability and a mature, highly disciplined infrastructure of government and private activities. The number of reactor years of safe operation has grown from over 2,400 to over 5,700, and the number of miles safely steamed on nuclear power by U.S. Navy vessels has increased from over 50 million to over 134 million since 1982.

John T. Conway, Chairman, Defense Nuclear Facilities Safety Board (DNFSB), said, "…the Naval nuclear propulsion program has set the standard for all of us who are committed to the safe design, construction, and operation of our nation's nuclear facilities. We are particularly mindful of the standards this program has set and maintained in the areas of nuclear safety, radiological protection, and environmental stewardship. Having the vision to set such standards, and the discipline to meet them for five decades, is an accomplishment unmatched in our nation's history…"[145]

The program has compiled an unparalleled record of success, including the following:

- Nuclear-powered warships have safely steamed over 134 million miles — equivalent to nearly 5,400 trips around the Earth.
- Naval Reactors are responsible today for 103 operating nuclear reactors. This is equivalent to the number of licensed commercial power reactors (104) in the United States. Over the years, they have accumulated over twice the operating experience of the U.S. commercial power industry. Naval reactor plants have accumulated over 5,100 reactor years of operation, compared to about 2,400 for the U.S. commercial industry. Their operating experience is about half that of the entire commercial power industry worldwide (their 5,100 reactor years compared to about 9,200 worldwide, including the United States).
- Naval Reactors' outstanding and fully public environmental record enables the U.S. ships to visit more than 150 ports

around the world — critical to the Nation's forward presence strategy and ability to project power.

The Navy nuclear ships and submarines have stayed on the leading edge of technology in regard to tactical speed, silencing, and reliability. The nuclear reactors and propulsion systems must be militarily capable and reliable in combat. The reactors must be safe for the environment, the public, and the crews that operate them. Because sailors have to live on the ships during operation, reactor compartments are designed to attenuate radiation levels outside of the reactor compartment to extremely low levels. The external surface radiation levels for the normal conditions of transportation in currently operating Navy vessels are but a fraction of the specified 200 millirem per hour. These plants can sustain battle shock and keep right on operating. They are resilient to accommodate years of frequent power changes.

Operated and maintained by a highly trained Navy crew, these small and uncomplicated pressurized water reactors are inherently safe and can respond to operation transients without the need for immediate operator action. The fission products are maintained within high integrity fuel elements that can withstand high shock loading. The reactor is so heavily shielded that a propulsion plant operator receives less radiation exposure from the reactor during a two month submerged patrol than he would receive from background radiation ashore.

The NNPP continues to advance reactor technology while exploring new energy conversion and reactor concepts that could better meet the unique requirements of naval nuclear propulsion plants. The program's objective is to make nuclear propulsion quieter, extend the life and efficiency of the nuclear core, further simplify operation of the reactor plant, reduce the already low life-cycle maintenance costs, and design propulsion plants with increased power output without increasing size or cost.

In addition to the military application of nuclear power, technology developed by the NNPP is the basis for civilian nuclear power around the world, such as:

- The uranium-dioxide fuel system, now the most widely used system in nuclear power;
- The design for large pressurized water reactor components and the cladding for large pressure vessels;
- Containment concepts and refueling techniques for power reactors;
- A system for preventing damage to a reactor core even if failures occur in the cooling system;
- The first successful method of radioactive decontamination of reactor plants;
- Zirconium, zirconium alloys, boron, and hafnium materials for cladding and reactor control use;
- Numerous computer programs widely used for design, safety, research, and testing;
- The first chemical cleaning process for nuclear plant steam generators;
- Ultrasonic inspection methods for evaluating the material status of the reactor vessel and major components;
- Nuclear fabrication standards, quality control requirements, and equipment specifications; and
- Development and publication of the Chart of the Nuclides, used worldwide for nuclear research and development work.

The Navy also shares with industry information from its research in the following areas:

- Corrosion and wear technology for components operating in high temperature, high pressure water;

- Pressurized water reactor heat transfer and fluid flow technology;
- Predicting performance of reactors in accident scenarios; and
- Numerical analysis of reactor design using digital computers.

From this research, more than 5,000 technical reports have been made available to the public and industry. However, the most important contribution to the civilian sector is the thousands of highly trained NNPP graduates who now play key roles in the operation and management of our civilian nuclear power plants.[146]

The nuclear propulsion plants in U.S. Navy ships, while differing in size and component arrangement, are rugged, compact, pressurized water reactors that are designed, constructed, and operated to exacting criteria. Fissioning of nuclear fuel produces the heat required, but the fissioning process also produces radiation. To protect the crew, the nuclear components of these plants are all housed in a section of the ship called the reactor compartment, usually buried deep in the hull and well protected.

Chapter 7, *Nuclear Reactors,* briefly describes two Light Water Reactors, the Boiling Water Reactor (BWR) and the Pressurized Water Reactor (PWR), which the Navy uses to power its ships and submarines. A particularly important advantage of the PWR is that the three water systems or loops never intermix; therefore, the risk of external radiation aboard ship is virtually zero. Steam drives both the turbine generators that supply the ship with electricity and the main propulsion turbines that drive the propeller(s). Because the three water loops are closed systems where water is re-circulated and renewed, there is no step in the generation of this power that requires the presence of air or oxygen. The advantage to the submarine fleet is obvious: It allows the ship to operate completely independent from the earth's atmosphere for an extended period of time.[147]

Naval reactors must undergo repeated power changes for ship maneuvering, unlike civilian counterparts that normally operate at a steady rate. Nuclear safety, radiation, shock, quieting, and operating performance dictate requirements; and operation of reactors in proximity to the crew necessitate particularly stringent safety standards. The internals of a Navy reactor remain inaccessible for inspection or replacement throughout a long core life, usually the life of the submarine. This is unlike a typical commercial nuclear reactor that is open for refueling every eighteen months.

Again, unlike commercial nuclear power plants, Navy reactors must be robust and resilient to withstand decades of rigorous operations at sea, subject to a ship's pitching and rolling, and rapidly changing demands for power, even under battle conditions. These conditions, combined with the harsh environment within a reactor plant, subjects components and materials to the long-term effects of irradiation, corrosion, high temperature, and pressure.[148] All of these necessitate an active, thorough, and far-sighted technology effort to verify reactor operation and enhance the reliability of operating these plants, as well as to ensure Naval nuclear propulsion technology provides the best options for future needs.

When the commercial nuclear industry was essentially abandoned in the 1970s, nuclear suppliers had no other work to absorb overhead costs and sustain a solid business base to compete for Navy nuclear work. As a result, competition is non-existent. Because the requirements for NNPP components are far more stringent than for civilian products, costly quality control and production procedures must be imposed. As a result, firms are less apt to compete in this environment. There is no civilian demand for quiet, compact, shock resistant nuclear propulsion systems that will keep skilled designers and production workers current. This is very different from aerospace, electronics, and ground vehicle industries that DoD supports.[149]

The NNPP has displayed to the world that nuclear power can be handled safely, with no adverse effects on the public or the environment. While others have stumbled with this challenging technology, the Navy program stands out in the private sector, as well as in the public sector for its *vision, discipline, and technical excellence.*

The U.S. Navy is the only organization in the world that designs, builds, operates, and *recycles* nuclear-powered ships, while exercising its environmental responsibilities throughout the ships' life cycle. Recycling involves defueling the reactor, inactivating the ship or submarine, removing the reactor compartment for land disposal, recycling the remainder of the vessel to the maximum extent practical, and disposing of the remaining non-recyclable materials. In the case of the USS Hyman G, Rickover, a submarine that operated for more than two decades helping keep Cold War foes in check and later supporting the ongoing Global War on Terrorism, recycling commenced upon its decommissioning in January 2009. It will be shredded into millions of pounds of steel and lead, and bundles of aluminum, brass, bronze, copper, and zinc. Spared the cutting torch, its reactor is destined for burial alongside 115 other Navy nuclear reactors.

The spent fuel removed from the nuclear warships constitutes less than 0.05% of all spent nuclear fuel in the United States today. It can be safely stored pending placement in a geologic repository, because it is designed to withstand combat conditions. The Navy has made over 900 container shipments of spent nuclear fuel since 1957 using specially designed, rugged containers, such as the M-140. To date there have been 90 nuclear warships recycled and their reactor compartments sent to DOE's Hanford Site.[150]

Over the past two decades, the U.S. Navy has downsized the United States Fleet, as naval weapons' systems reached obsolescence, and because the ending of the Cold War* reduced military force requirements. (*Note: The Union of Soviet Socialist Republics [USSR) began to crumble in 1989 and collapsed in 1991, which marked the end of the Cold War.) The Los Angeles- and Ohio-class submarines, and all nine nuclear-powered cruisers, as well, have been decommissioned and removed from active service. However, some members of Congress believe the lack of nuclear-powered surface combatants to be a serious naval warfighting deficiency.[151] Therefore, nuclear power is being reconsidered for medium-size surface combatants — the next-generation cruiser, the CG(X), is a prime candidate.

Although the Navy considered disposing of reactor compartments by sinking them in the oceans, they chose the option of land burial, judging that the costs and potential for adverse environmental effects favored the latter option. Other DOE burial sites were considered, but Hanford in the state of Washington was chosen. Naval Station Bremerton, largely because of its proximity to Hanford, was chosen as the yard that would "scrap" all nuclear-powered cruisers and many of the submarines. Since 1986, the Navy has disposed of submarine reactor compartments at the Hanford Site and, beginning in 1999, reactor compartments from deactivated cruisers have been disposed there, as well. Through September 2007, 117 reactor compartments have been taken there. Except for the period 2005-2007, when only three reactor compartments were shipped, the average was six to eight compartments per year.

As part of the deactivation process and prior to transfer of the crew, the reactors are *defueled*. "Defueling" removes the nuclear fuel from the reactor pressure vessel, which removes most of the radioactivity from the reactor plant. Fluids are drained and pipes are sealed, and the spent fuel is sent to the Idaho National Laboratory (INL) for storage. (Defueling is accomplished at designated shipyards that perform reactor work for the Navy.) The reactor compartment is then cut from the rest of the ship or boat, sealed, and removed. Preparation time from start to "ready for transport" is from six to eight months for the submarines. (More time was required for the larger reactor on the cruiser U.S.S. Long Beach, because its reactor weighed half again as much [about 2,250 tons] as reactors on the other cruisers and submarines [about 1,400 tons]).

Historically, ships no longer needed for service in the active fleet or the reserves and not scheduled to be scrapped, may be placed in protective storage for an indefinite period of time without harming the environment, followed by permanent disposal, recharging, or recycling. If placed in floating storage, every 15 years the ship will be taken out of the water for inspection and repainting of the hull to insure continued safe water storage. The preferred land burial of the entire defueled reactor compartment is at DOE's Low-Level Waste Burial Grounds at Hanford, where it is still permitted. However, this protective storage does not provide a permanent solution for the reactor compartment.

Regarding safety of transport, the Navy must comply with Department of Transportation (DOT) regulations when shipping reactor compartments. To protect workers involved and the general public, radiation levels must not exceed DOT-specified limits. Investigations by concerned parties have given the Navy excellent marks. In point of fact, radiation levels of reactor compartments in shipment are virtually undetectable a few yards removed.

Before a ship is taken out of service, the spent fuel is removed from the reactor pressure vessel of the ship in a process called defueling, which removes all the fuel and most of the radioactivity from the reactor plant of the ships. The fuel removed from the decommissioned ships will be handled in the same manner as that removed from ships that are being refueled and returned to service. Unlike the low level radioactive materials in defueled reactor plants, the Nuclear Waste Policy Act of 1982, as amended, requires disposal of spent fuel in a deep geological repository. The United States has selected Yucca Mountain as its deep geological repository, and the Navy was the "driver."

Prior to disposal, the reactor pressure vessel, radioactive piping systems, and the reactor compartment disposal package is drained and sealed. This acts as a containment structure for the radioactive atoms and delays the time when any radioactivity might be released to the environment should the packaging deteriorate. This is very important, because radioactivity decays with time and would cause no harm to the environment. Over 99.9% of these atoms are an integral part of the metal, and they are chemically like ordinary iron, nickel, or other metal atoms. These radioactive atoms are only released from the metal because of the slow process of corrosion. The remaining 0.1%, which is corrosion and wear products, decays prior to corrosive penetration of the containment structures.

The Hanford Site is used for disposal of radioactive waste as stated above. The pre-Los Angeles-class submarine reactor components were placed there at the 218-E-12B burial ground in the 200 East Area. The disposal of the reactor compartment from the cruisers, and the Los Angeles- and Ohio-class submarines would be consistent with the pre Los Angeles-class submarine reactor components disposal program. The land required for the burial of 100 reactor components would be ten acres. An estimated cost for land burial of the reactor components is $10.2M for each Los Angeles-class submarine reactor, $12.8M for each Ohio-class submarine reactor component, and $40M for each cruiser reactor component. The estimated total shipyard occupational exposure to prepare the reactor compartment disposal packages (listed by ship type and class) is the following:

- For each Los Angeles-class submarine package, 13 rem,* approximately 0.005 additional latent cancer fatalities;

- For each Ohio-class submarine package, 14 rem, approximately 0.006 additional latent cancer fatality for each; and

- For each cruiser package, 25 rem, approximately 0.01 additional latent cancer fatality.[152]

  (*Note: rem is *roentgen equivalent man*, a radioactivity measurement used to determine equivalent dose depending on the type of radiation.)

Even after the vessels are shut down and the nuclear fuel is removed, the propulsion plants of nuclear-powered ships remain a source of radiation. Defueling removes all fission products since the fuel is designed, built, and tested to ensure that fuel will contain the fission products. Over 99.9% of the radioactive material that remains after the fuel is removed is an integral part of the structural alloys forming the plant components. The radioactivity was created by neutron irradiation of the iron and alloying elements in the metal components during operation of the plant. The remaining 0.1% is radioactive corrosion and wear products that have been circulated by reactor coolant. They have become radioactive from exposure to neutrons in the reactor core and then deposited on piping system internals.

Radioactivity is created during fission, because some of these fission products are highly radioactive when they are formed. Most of the radioactivity produced by nuclear fuel is in the fission products. The uranium fuel in naval nuclear propulsion reactor cores uses highly corrosion-resistant and highly radiation-resistant fuel and cladding. As a result, the fuel is very strong and has very high integrity. These fuel rods are designed, built, and tested to ensure that the fuel construction will contain and hold the radioactive fission products. Naval fuel contains fission products, but there is no product released during normal operation.

Most of the neutrons produced while the naval reactor is operating are absorbed by the atoms within the fuel and continue the chain reaction.[153] However, some of the neutrons escape and are absorbed in the metal structure that supports the fuel or in the walls of the reactor pressure vessel. Trace amounts of corrosion and wear products are carried by reactor coolant from reactor plant through the piping systems, where a portion of the radioactivity is removed by a purification system.

Protection of the environment was a high priority in the NNPP long before it became a popular public issue. From the onset, the program recognized that the environmental aspects of U.S. nuclear-powered ships and their operations would be key to their acceptance in ports both at home and abroad. Environmental releases, both airborne and waterborne, are strictly controlled. As a result, the annual releases of long-lived gamma radioactivity are comparable to the annual releases from a typical commercial nuclear reactor operating under the auspices of the Nuclear Regulatory Commission (NRC).

Through the entire history of the program, which includes more than 5,700 reactor-years of operation and over 134 million miles steamed on nuclear power, there has never been a reactor accident. There has been no release of radioactivity that has hurt human life, had an adverse effect on marine life, or had a significant effect on the environment. The program's standards and record surpass those of any other national or international nuclear program. R. L. Seale, Chairman, NRC Advisory Committee on Reactor Safeguards (ACRS), observed that through the years, he had been impressed by the conservative, robust Naval designs; the high standards for materials, fabrication, and installation; and the firm technical discipline used in the NNPP in selecting, training, and qualifying competent officers and enlisted personnel to operate plants.

The NNPP has an environmental monitoring program at each of its major installations and facilities, including shipyards and homeports. It consists of analyzing water, sediment, air, and marine samples for radioactivity to verify that operations have not had a significant effect on the radioactivity in the environment. Surveys by the Environmental Protection Agency (EPA) and state and local governments confirm that U.S. Navy nuclear-powered ships and support facilities have had no significant radiological effects on the environment.[154]

Ensuring proper environmental performance has always been a priority at Department of Energy facilities. The EPA holds regular inspections of the program's laboratory and prototype sites. These are in accordance with the Clean Air Act, the Resource Conservation and Recovery Act (RCRA), and the Clean Water Act. Also, none of these sites qualifies for inclusion on the EPA's National Priority List for cleanup under the Comprehensive Environmental Response, Compensation, and Liability Act (CERCLA) or Superfund.[155]

In August 1998, the Naval Nuclear Propulsion Program (NNPP) celebrated its 50th year of service to the United States, providing safe, effective, and environmentally sound nuclear propulsion to our Naval ships. In the Congressional Record, Proceedings and Debates of the 105th Congress, second session, vol. 144, no. 106, Washington, Friday, July 31, 1998, "The Senate Resolved That: (1) the Senate commends the past and present personnel of the Naval Nuclear Propulsion Program for the technical excellence, accomplishment, and oversight demonstrated in the program and congratulates those personnel for the 50 years of exemplary service that has been provided to the United

States through the program; and (2) it is the sense of the Senate that the Naval Nuclear Propulsion Program should be continued into the next millennium to provide exemplary technical accomplishment in, and oversight of, Naval nuclear propulsion plants and to continue to be a model of technical excellence in the United States and the world."

Accolades by the following dignitaries reflect the respect and gratitude of the people of the United States for their Navy and its Naval Nuclear Propulsion Program.

Former President Bill Clinton: "…The Naval Nuclear Propulsion Program embodies unsurpassed engineering and sustained excellence that few technical programs in or out of government can claim. In every area of performance, standards, safety, and environmental care, the Naval Nuclear Propulsion Program has excelled."[156]

Then Secretary of Defense, William A. Cohen: "…Since 1948, the Program, founded by Admiral Rickover, has served as a shining example of the excellence and efficiency we are working to achieve throughout the Department of Defense. Nuclear propulsion has enabled the Navy to make vital and continuing contributions to our national security around the globe. The reliability and endurance of the nuclear-powered fleet have been, and remain today, pivotal to providing strategic deterrence, forward presence, and freedom of the seas…"[157]

Former Secretary of Energy, Bill Richardson: "…I am aware of the enormous contribution your technological accomplishments have made to the nation. For 50 years the Naval Reactors Program has delivered the finest performance in nuclear technology development, reactor operations, safety, and environmental protection…"[158]

Also, General Henry Shelton, then Chairman of the Joint Chiefs of Staff on the Program's 50th anniversary, said, "…Over the last half-century, Naval nuclear reactors have steamed over 110 million miles with an unmatched, absolutely flawless record of safety and performance. Today, nuclear-powered aircraft carriers reign as the centerpiece of America's strategy of forward presence, and nuclear-powered submarines remain a crown jewel of our nation's defense arsenal..."[159]

Commissioned in 1955, the USS Nantilus signaled the beginning of the nuclear-powered submarine fleet. Its combat potential was demonstrated when she sailed submerged from New London, Connecticut, to San Juan, Puerto Rico, a distance of 1,350 miles in 84 hours; its record-breaking cruising speed while submerged was more than 20 knots. Subsequent nuclear submarine accomplishments include the following series of significant events. In August 1958, the Nautilus made the first undersea transit of the North Pole, from Point Barrow, Alaska, between Norway and Greenland. Later, the same month, the USS Skate, another nuclear-powered submarine launched in 1957, together with the USS Seawolf and USS Swordfish, reached the North Pole during an exploration trip. The Seawolf set an endurance record for underwater operation of 60 days between August 7 and October 6, 1958. In 1956, the USS Skipjack was launched, combining nuclear propulsion with the blimp shaped hull and a single propeller. Advanced versions of this submarine, known as the Thresher-class, were placed in operation in the early 1960s. On April 10, 1963, the USS Thresher was lost with 129 men aboard during deep-diving tests in the Atlantic about 200 miles east of Boston. This tragedy resulted in innovations in submarine design and undersea rescue technology.[160]

In 1960, the first submarines incorporating a battery of Polaris missiles were built in the United States. In the late 1960s, the longer range Poseidon missile, which was capable of carrying up to ten nuclear warheads, replaced the Polaris missile on some submarines. In the early 1970s, the United States accelerated the Trident I system, a successor to the Polaris-Poseidon. The Trident I system included the new Ohio-class nuclear submarines equipped with 24 launching tubes, each tube containing an ICBM with a range of 4,600 miles. The first vessel of this class was the USS Ohio, which was launched and commissioned in 1981. By 1988, the United States had 132 submarines in operation, almost all of them nuclear-powered. Most of them have a reactor designed to provide propulsion for at least 400,000 miles without refueling. Nuclear power offered a way to drive the submerged submarine at high speeds without concern for fuel consumption, have the ability to operate fully-capable sensors and weapons systems during extended deployments, and maintain a safe and comfortable living environment for the crew.

The latest generation submarine is the Virginia-class attack submarine, a joint venture shared by General Dynamics Electric Boat and Northrop Grumman Newport News. The USS Virginia (SSN 774), the first major combatant designed since the end of the Cold War, is the first to use modern, state-of-the-art modular manufacturing techniques. Replacing the Seawolf-class, the Virginia is a smaller, cheaper, multi-mission, stealth submarine for open-ocean anti-submarine warfare and littoral (along the shore, shallow water) operations. Commissioned in October 2004, the Virginia completed its deployment in support of the Global War on Terrorism in November 2005.[161]

It is important to recognize the aircraft carriers, as well as the submarines, because the impact of nuclear power on naval aviation was an equally dramatic leap forward in capability. No longer tied to slow at-sea supply lines for propulsion fuel, the aircraft carrier with its embarked air wing could deploy rapidly over great distances to respond to crises effectively and efficiently. The nuclear-powered aircraft carrier (CVN) would arrive on station earlier with its air wing ready to fight, and it could remain on station sustaining the fight far longer than its fossil-fueled predecessor. The CVN has nearly unlimited propulsion endurance and twice the aviation fuel storage of conventional carriers of comparable displacement.

Today, the 103 nuclear reactors in the U.S. Navy power roughly 40% of its major combatant ships. All aircraft carriers and submarines, including all of the ballistic missile submarines, are nuclear-powered. The few cruisers and destroyers that were powered by nuclear energy have been decommissioned. Virtually undetectable while submerged, the strategic ballistic missile submarines are the least vulnerable leg of the U.S. strategic deterrent force. The 14 Trident Ballistic Missile submarines armed with a total of 336 Trident missiles carry about 50% of the United States strategic nuclear warheads. At 560 feet in length and 18,700 tons displacement, the Trident is the largest U.S. nuclear-powered submarine.[162]

The attack submarines, forward deployed, alone and unsupported, can exert their influence throughout the world's oceans and seas. They protect vital commercial sea-lanes, provide protection for the U.S. surface warships, and create tactical uncertainty for an enemy, who must tie up fleet units in defensive roles. These submarines operate undetected in all waters throughout the world even under ice in the Arctic Ocean. With their cruise missiles, targets ashore are not immune to their firepower, even while submerged. Today's active attack submarine force consists of 47 Los Angeles SSN 688 - class submarines, three Seawolf SSN 21 – class, and two Virginia SSN 774 – class that are the most advanced attack submarines in the world.[163]

The nuclear-powered aircraft carriers can transit at high speed, alone or escorted, to the scene of the crisis fully ready to launch their awesome firepower. They can sustain that presence with tactical mobility and flexibility without refueling or immediate replenishment of combat consumables. Today's nuclear-powered carrier fleet consists of the USS

Enterprise CVN 65, the first-nuclear-powered aircraft carrier, and nine Nimitz CVN 68 - class supercarriers that are the largest warships in the world. The tenth and last of the class, the George H. W. Bush (CVN 77), was christened at Northrop Grumman Newport News in October 2006 and commissioned on January 10, 2009.[164]

Relying only on the NNPP's 60 years of demonstrated public record of safe operations and management practices, foreign governments grant U.S. nuclear-powered warships access to their ports. The Navy's reputation for operation and maintaining nuclear-powered warships in a manner that poses no risk to the public or to the environment is central to U.S. nuclear-powered warship access to foreign ports. These same considerations are important to the continued acceptance by state and local governments, and to the public of the United States. In contrast, it is widely known that the former Soviet Union naval nuclear propulsion program has had significant problems including reactor safety, personnel radiation exposure, environmental protection, and radioactive waste management. Because of their operating practices, nuclear-powered warships of the former Soviet Union are rarely seen in foreign ports.

The United States Navy Nuclear Fleet Set the Standard.

CHAPTER 6

France

A little experience is worth a great deal of reading.
Thomas Jefferson[165]

Frenchman John Ardagh in his book, *FRANCE in the New Century: Portrait in a Changing Society*, shares his views regarding France's economic, social, and cultural conditions as they exist in the 21st Century.[166] Ardagh explains that when the explosive price of crude oil hit in the late 1960s and early 1970s, France was importing over 75% of her energy needs: against an European Economic Community (EEC) average of 55%. Valery Giscard d' Estaing's government reacted vigorously to the energy crisis with a three-point strategy:[167]

1. Enact strong conservation measures
2. Search for alternative energy sources
3. Expand nuclear power generation capability.

It is not difficult to understand why the French Government has put such an emphasis on nuclear for so long. Exploding her first atomic bomb in 1960, she joined the United States and Russia to become what was known as the nuclear power triad. Initially, Charles de Gaulle pursued a nationalist "go-it-alone" policy, spurning American technology and relying on home-produced gas-cooled reactors. Surprisingly, in the late 1950s he opened the door for evaluation of U.S. nuclear reactor technologies. Then in 1969 Georges Jean Raymond Pompidou, de Gaulle's successor, made an historic turn-around; he switched to an American light-water system, based on enriched uranium. Pompidou felt it was the only reasonable, economic choice, even though he had placed some dependence on U.S. technology. Under license from Westinghouse, Framatome, a French nuclear firm, began to build pressurized-water reactors (PWR).[168] Cost was the most important factor in the decision by the French Government to standardize her nuclear power plants using PWRs.

The Framatome-Westinghouse transaction placed France in an excellent position when the energy crisis occurred, and she pushed ahead with a civil *nucleaire programme*, which became far bigger than her military nuclear acquisitions. Then without relying on American technology, the French became the pioneers of the fast-breeder reactors (FBR). In 1977, work began on the Superphenix, a multi-national French-led project for a 1,200 megawatt fast breeder utilizing uranium and plutonium. It had its problems, but it catapulted France to the forefront in nuclear technology. Today, France remains ahead of other countries both in Europe and abroad in nuclear power development for industrial applications.

The civil *nucleaire programme* has come under attack from environmentalists, but they have made less impact in France than in other countries. After the TMI accident[169] in 1979, officials of the *Electricite de France* (EdF), the national electricity provider, claimed that because France had different reactor designs, she was less likely to experience a similar accident. As a result, Valery Giscard d'Estaing ordered a speedup of nuclear construction — at the time a decision unthinkable in any other country. Officially the view was that if national independence and high living standards were to be maintained, France had no other alternative.

By 1981, France was completing five or six new reactors a year — building as many nuclear plants in two years as Britain had in 30. Currently, there are 58 nuclear power stations in operation, producing 79% of French electricity. The United States is at 20% (104 reactors), Japan 30% (55 reactors), Russia 16% (31 reactors), United Kingdom 20% (23 reactors), South Korea 45% (20 reactors), and Canada 15% (18 reactors). In 2006, 16 countries relied on nuclear power to generate more than 25% of their electricity. Twelve generate more electricity using nuclear than any other energy source: France, of course, leads the list with 79%, followed by Lithuania, Slovakia, Belgium, Sweden, Ukraine, Bulgaria, Armenia, Slovenia, South Korea, Hungary, and Switzerland with 37%. The list is interesting. The countries range from sophisticated economies to developing nations. In September 2007, the World Nuclear Association (WNA) reports, "Nearly 20 countries are actively considering embarking upon nuclear power programs."[170]

By standardizing her nuclear power plant designs, France tightened the links amongst research and development (R&D), industry, and oversight authorities.[171] Each party is well aware of its responsibilities and completes its tasks with the care and professionalism required in the nuclear industry. She has also mastered the entire nuclear fuel cycle from extraction of natural uranium to enrichment and manufacturing techniques to reprocessing and recycling spent fuel.

France has invested more than FRF 400 billion 1993 currency (about $87 billion) in her nuclear program over the past 30 years and has created more than 100,000 jobs. Today, as her nuclear activity has shifted somewhat to operation and maintenance, about 55,000 people are employed in the nuclear industry. Other businesses have benefited from the nuclear program, mainly health sciences, radioactive isotopes for medical diagnosis and treatment, microelectronics, and agribusiness.[172] Although French nuclear equipment has been costly to install, electricity can now be produced cheaply on a large scale,[173] so much so that France is now exporting 16% of her nuclear-generated energy.[174] From being an importer in the 1970s, France is now Europe's largest exporter of electricity. Italy, having no operating nuclear power plants, is the largest electricity importer in Europe, and most of it comes from France.[175] ("Italy announced May 22, 2008, that within five years it planned to resume building nuclear energy plants, two decades after a public referendum resoundingly banned nuclear power and deactivated all its reactors.")[176] The United Kingdom is also a major customer for French electricity.

Extraordinary communication efforts have been expended by French nuclear corporations that invite the public to visit their facilities, develop information adapted to the populations living in the vicinity of the plants, to teachers, medical corps and other health personnel, elected officials, and oft-times even the public at large. These corporate publicity drives have been supplementing traditional communication since 1992 — in other words *France educates the public. The United States does not.* France also has shown that the benefits from being largely nuclear are numerous and the public will accept nuclear power if the facts are known.

The Commissariat a l'Energie Atomique (CEA) is the French Atomic Energy Commission, a government-funded technological research organization. Created in 1945, CEA is a world leader in research, development, and innovation. Active in both national defense and nuclear energy, its defense responsibilities include nuclear propulsion and warheads in energy it leads in new technologies, future systems, and nuclear waste. For over 50 years, CEA has worked hand-in-hand with *Electricite de France* (EdF), which was created the following year in March 1946. The EdF is a state-owned utility that is responsible for generation, transmission, distribution, importation, and exportation of electricity in France. From the outset the long-term priority of EdF has been cost-effectiveness.

In 2006, the French Government asked EdF and AREVA to build a next-generation nuclear reactor, the European Pressurized Reactor (EPR) at the Flamanville Nuclear Power Plant located on the Cotentin Peninsula. Currently, the plant houses two PWRs that came into service in the late 1980s.[177] (The EPR will be discussed in Chapter 7, Reactors.)

AREVA, a public multinational industrial conglomerate, was created in September 2001 by the merger of CEA Industrie, Framatome, and Cogema. It is the world leader in nuclear energy and the world's most experienced nuclear industry services company. Three main companies form the core of AREVA. They are the following:

- AREVA NP (formerly Framatome ANP) — Designs, develops, and builds nuclear reactors;
- AREVA NC (formerly Cogema) — Specializes in the nuclear fuel cycle from mining to waste disposal; and
- AREVA T&D — Power transmission and distribution.

The AREVA family is part of the Global Nuclear Energy Partnership (GNEP) alliance. Other members are the Japan Atomic Energy Agency (JAEA), Washington Group International, and Babcock & Wilcox (BWX). The GNEP plan is to reprocess spent nuclear fuel rendering the plutonium recovered usable as nuclear fuel but not for nuclear weapons.

AREVA's experience dates back to 1954, when it built a spent fuel handling facility for the French Atomic Energy Commission (CEA). Today, its nuclear services operations are worldwide. AREVA possesses extensive experience in *front-end* activities, such as mining, conversion, and enrichment. *Back-end* fuel cycle problems were resolved, using highly developed recycling and fuel treatment technologies that are cleaner, more efficient, less waste intensive, and more proliferation resistant. Thanks to France's steady policy of reprocessing and recycling, she has accumulated the world's greatest depth of experience in commercial recycling and vitrification of liquid high-level waste, as well as in fabrication of mixed oxide fuel (MOX) from recycled plutonium.

Blessed with having large uranium deposits in her backyard, France was the leading European producer of uranium for decades; output was 6,500 – 7,000 tons per year, approximately 20% of the world's production. But, in May 2001, the last producing mine was shut down. However, AREVA's explorations worldwide, including Indonesia, Canada, Australia, and the United States, have led to the discovery of deposits that will ensure a long-term supply of uranium. In the United States, AREVA is mining by the underground method in Texas and Wyoming where available reserves exceed several million pounds of uranium. In Canada, deposits being developed in Saskatchewan's Cluff Lake Site offer the highest assays of uranium in the world, 28 to 210 kilograms of uranium per metric ton of ore, which is 10 to 100 times richer than previously mined deposits. Of the nuclear reactors in operation or under construction, 90% are fueled with uranium that has been enriched between 3% and 5%, whereas natural uranium contains only 0.7% U_{235}.

AREVA NP partnered with BWX in the United States to design and build the Fuel Master system to help the United States double the storage space in the nation's spent fuel pools. The automated, robotic process produces storage canisters containing twice the number of rods as conventional assemblies. By using Fuel Master, the United States can minimize the need for costly storage upgrades while increasing the flexibility of the nation's existing facilities.

AREVA NC has site remediation and environmental cleanup capabilities that are designed to be compatible with International Standards Organization (ISO) 14001 standards. Activities involved in site remediation are dismantling of installations, decontamination of soil, and disposal of waste products in strict accordance with environmental and radiological regulations. The sites are monitored during and after remediation to ensure radioactivity levels are below regulatory limits. They are also stabilized and landscaped according to the type of mine, be it open, underground, or lixiviation, with emphasis given to the wishes of local residents and other land-user groups.

AREVA T&D supplies products, systems, and services for transmission and distribution of electricity. All aspects, including automation of the processes, of the total T&D system necessary to regulate, switch, transform, and dispatch electric current in the networks that connect the generating plant to the user fall under the auspices of the T&D division.

In addition to EdF and AREVA, ANDRA (The French National Agency for Radioactive Waste Management) is also a major component in the structure of the nuclear industry in France. Created in 1991 and operating independently of waste producers, Andra is responsible for the long-term management of nuclear waste produced in France. Quality assurance, safety, and environmental preservation are the core of its responsibilities.

Throughout the nuclear industry, reliability, maintainability, and safety must rely on technology. Reliability combined with preventive maintenance dictate the availability of the operational systems and processes. Experience plays the primary role in this scenario. Proven technologies are invaluable, because they tend to eliminate long and costly component testing, full-scale testing of equipment, and operational testing of new systems. With nearly 50 years of uninterrupted experience with plant design, nuclear operations, and waste management projects, AREVA brings to the table utmost concern for reliability, cost-effectiveness, and safety in all nuclear fuel cycle operations. *France's positive approach to nuclear energy policy, supportive role toward the industry, and AREVA's successful management and operation serve as models to emulate in planning for and implementing new nuclear initiatives.*

National support has been key to continued success of the nuclear power enterprise in France. Bear in mind that Charles de Gaulle founded the Fifth Republic in 1958, and this government has lasted nearly 50 years, across five presidencies — (1959-69, his own); George Pompidou (1969-74; died in office); Valery Giscard d'Estaing (1974-81); Francois Mitterrand (1981-95); and Jacques Chirac (1995-2007). Elected in May, the current Prime Minister is Nicolas Sarkozy, who has promised closer ties with the United States. He is in favor of development of renewable energies and renewal of the French nuclear fleet. This has positive implications for future French nuclear initiatives.

During much of the Fifth Republic administration, ethnic movements grew increasingly Leftist in a climate of increasing Gaullist nationalism. The Socialists became reticent, and this put the Left in a dilemma when it took power with the election of Mitterrand in 1981. The party was anti-nuclear, but Mitterrand was equivocal. He and his government faced the facts. They knew that to halt nuclear expansion could be costly and would place over 100,000 jobs at risk, so they decreed a slowdown, halting work on five of the 14 plants then under construction. The decision was taken primarily for political reasons, but the rate of growth of consumption was less than anticipated as well. Even though the Left did slow the *nucleaire programme*, nuclear capacity more than doubled between 1981 and 1989; from 19 to 48 units on line during the 8-year span. From then until 2002, only 10 units were added.

In 2005, EdF announced plans to replace current nuclear plants as they reach the end of their licensed life starting around 2020. This decision confirms nuclear power as the energy source of choice for future power plants. Implementation will require new construction of one large (1,600 megawatts electric [MWe]) unit each year for 40 years in order to replace the 58 operating today.

Clearly, France has made a major effort to reduce its dependence on imported energy. Her self-sufficiency rose from 22% in 1973, to 53% in 1985, but then fell to 46% in 2003. Electrical generation by nuclear power rose from 1% in 1973, to 30% in 1981, to 76% in 1998, to 77% in 2000, to 78% in 2003. Dependence on imported energy has fallen from 78% to 54%, while oil imports have increased (measured in thousands of barrels per day) from 1,765 in 1993 to 1,984 in 2003, an increase of 12%. France's energy production and consumption have both been increasing over the past ten years.[178]

Although France had invested some 400 billion francs (~$73.1 billion U.S. [1993 dollars]) in her nuclear technology by 1997[179], she postponed replacing her aging stations. By 2000, nuclear construction was at a virtual standstill under Dominique Voynet, Minister of the Environment and Regional Planning from 1997 until she resigned in July 2001. The EdF no longer had a free hand in its day-to-day running of nuclear power. Before, the government let it assess its risks, make its decisions, then offer explanations to the public. (A founding member of The Greens in France, Voynet ran as the Green candidate for President in 2006.)

Even though Lionel Jospin, Prime Minister from 1997 – 2002, had stated that the nuclear industry is an important asset, his party was deeply divided, just as was Mitterrand's. In France, the Communists have always been pro-nuclear, for patriotic and employment reasons, as has the *General Confederation of Labor/Force Ouvriere* (CGt-FO) France's trade federation union. Its leader (1989 to 2003) at the time, Marc Blondel stated, "I'm pro-nuclear; I know what it is for miners to end up at 45 years with silicosis. The French public is still in favour of nuclear power."

It is clear that France relies heavily on her nuclear industry, which is almost as valuable to France as her *automobiles* and *aeroplanes*. The French, proud and materialistic, are not likely to let anyone or anything erode her *nucleaire programme*. Development of her nuclear power industry offers the following *economic advantages*:

- Considerable reduction in currency spent overseas;
- Nuclear energy reduces fossil fuel imports;
- Electricity exports, 68.3 billion kWh in 2005, is a major positive factor in trade balance;
- Low electric power prices to the customer, since the price of the kWh; produced from fossil fuels is 20% higher than that of the nuclear kWh; and
- Direct (100,000) and indirect jobs created are considered a national resource.

France has built a nuclear industry that competes worldwide.

CHAPTER 7

Nuclear Reactors

That one hundred and fifty lawyers should do business together ought not to be expected.
Thomas Jefferson to U.S. Congress[180]

Dr. Paul Gray, past president of Massachusetts Institute of Technology, made these remarks nearly 20 years ago, and still no action has been taken.[181]

Steam still generates most of the world's electricity. Coal, gas, oil, and nuclear power are used to turn water into steam to drive turbines that produce electricity. Huge quantities of gas and imported oil are consumed for other energy requirements, such as transportation. Burning fossil fuels is expensive and deteriorating the environment. Oil accounts for over half of our entire balance of payments deficit. More than a billion dollars per week is spent on foreign oil imports. It travels up the chimney, out the tailpipe of the cars, and into the atmosphere.

There is a cleaner, more economical, and safer way to generate electricity, and that is with nuclear power.

Nuclear reactors are grouped according to *generation* to provide a sense of the sequence in which they evolved. The first developed, Generation I reactors, have been retired from commercial operation. Generations II and III are in operation around the world. Generation IV reactors, by and large, are theoretical designs not expected to be operational before 2020. The most common reactors operating in the United States are classified *light water reactors* (LWR), a collective term that includes both the boiling water reactor (BWR) and the pressurized water reactor (PWR). These reactors cannot explode, and the reactor fuels in their slightly enriched state *can not* be used to make nuclear bombs. Countries with the largest number of reactors are the United States, France, and Japan. Those with the most under construction are India (4), Japan (4), Russia (3), and China (2). Currently, twelfth in the world in production of electricity by nuclear power, China has an ambitious program that would put it ahead of the United States, the current leader, by the year 2030.

Of interest in the consideration of recycling nuclear waste are those Generations II and III reactors that can be converted or modified to burn reprocessed mixed oxide (MOX) fuel and the new Generation IV designs. Nuclear reactors of interest, several of which will be discussed in some detail, are as follows:

Generation I Reactors (1946-1965)

All Generation I reactors in the United States are now out of service, but three reactor plants of historical interest are the following:

- Shippingsport
  The first large-scale nuclear power plant in the world began operating in Shippingport, Pennsylvania, December 2,

1957 — exactly 15 years after Enrico Fermi demonstrated the first sustained nuclear reaction.[182] Shippingport Atomic Power Station was built on the Ohio River near Pittsburg by the Duquesne Light Company. Its government-owned reactor was designed by the Westinghouse Electric Corporation in cooperation with the Division of Naval Reactors of the Atomic Energy Commission (AEC). As part of his "Atoms for Peace" program, ground was broken in 1954 during dedication ceremonies by President Dwight D. Eisenhower, who also opened it on May 26, 1958. Shippingport's nuclear power plant was retired in 1982, and the Congress assigned to the Department of Energy (DOE) responsibility for decontamination and decommissioning (D&D), the first D&D of a commercial reactor in the United States

- Dresden

  The Dresden Nuclear Power Plant, located in Morris, Illinois, was the first privately financed nuclear power plant built in the United States. Dresden 1 was activated in 1960 and retired in 1978.

- Fermi

  A third Generation I plant was the Fermi Nuclear Generating Station, located on the shore of Lake Erie near Monroe, Michigan. Fermi 1 was activated in 1963 and retired in 1972.

Generation II Commercial Power Reactors (1965-1995)

Generation II reactors constitute the majority of reactors operating today, safe and reliable, but are being superseded by better designs.

- Boiling Water Reactor (BWR)

  The BWR is comparable to the Pressurized Water Reactor (PWR). In the BWR's direct cycle, the water boils in contact with the fuel and goes directly to the turbine. The primary and secondary water loops are not separate systems.

- Pressurized Water Reactor (PWR)

  The PWR operates at a sufficiently high pressure that the reactor coolant does not boil. The water is heated, pumped to an interior generator (heat exchanger) where secondary water is boiled. (Today, the PWR is the most often installed nuclear power source in the world. These reactors equip the bulk of the French nuclear parks as well.)

- Heavy Water Reactor (HWR)

  The HWR was developed in Canada. Heavy water acts as the moderator and, because its properties are similar to ordinary water, is used as the coolant as well.

- Canada Deuterium Uranium Reactor (CANDU)

  A PWR, the CANDU is the only power reactor currently in use in Canada. Using natural uranium as fuel, the CANDU is popular in countries lacking enrichment plants.[183]

- Advanced Gas-cooled Reactor (AGR)

  The AGR is a second-generation British-developed design. Graphite acts as the moderator and carbon dioxide as coolant. It operates at higher temperatures for improved efficiency and uses enriched uranium fuel requiring less frequent refueling.

Generation III Advanced Light Water Reactors (1995-2010)

Many Generation III reactors incorporate passive safety features, lower capital cost, longer life, and higher efficiency.

Advanced Boiling Water Reactors (ABWR)

- Advanced Boiling Water Reactor (ABWR) (General Electric [GE], Toshiba, and Hitachi)

  The ABWR was certified by the Nuclear Regulatory Commission (NRC) in May 1997. The ABWR is a direct cycle BWR that reflects 50 years of evolution from GE's original concept. Four units are operational in Japan and 12 more are under construction or planned in the Far East (Taiwan and Japan).

- Siedewasser Reactor-1000 (SWR-1000) (AREVA NP)

  The AWR-1000 is under contract with German utilities. The AREVA NP developed a design that uses passive systems to decrease the risk of human error.

- High Conversion BWR (HCBWR)

  Research is ongoing to develop an economically competitive, passively safe, and proliferation resistant high conversion BWR for the specific purpose of burning existing stocks of plutonium, while converting the fertile thorium to U_{233} (a fissile isotope of uranium).

- System 80+ (ABB Combustion Engineering)

  Certified by NRC in May 1997, System 80+ is an advanced PWR. It provided the basis for the APR 1400. An interesting feature of System 80+ is that it can burn plutonium fuel.

- AP600 (Westinghouse BNFL)

  The AP600 was certified by NRC in December 1999.

Generation III+ Evolutionary Designs Offering Improved Economics (2010-2030)

Advanced Pressurized Water Reactors (APWR)

These reactors evolved from existing PWRs. Design innovations include passive safety, cooling by gravity flow, and use of existing technologies to reduce cost.

- AP1000 (Westinghouse BNFL)
  The AP1000 is based on the AP 600 and was certified by NRC in January 2006. An amendment is under review.
- APR-1400 (Westinghouse BNFL)
  The APR-1400 design evolved from the System 80+. The reactor was developed in Korea as the KNGR (Korean Next Generation Reactor) for deployment and export.
- European Pressurized Water Reactor (EPR) (AREVA NP [Framatome ANP]) A PWR, the EPR is expected to be the replacement reactor for France's electric utility system. Certification is under review (2011).
- Economic Simplified BWR (ESBWR) (General Electric)
  The ESBWR, based on the ABWR, uses natural circulation, incorporates passive safety systems, and relies on economies of scale in construction. It is undergoing NRC design certification. Certification is expected in 2010.
- Canada Deuterium Uranium-Advanced CANDU Reactor (CANDU-ACR) (Atomic Energy of Canada Limited [AECL]). Also known as ACR700 or ACR1000, the CANDU-ACR is a Modified Pressurized Heavy Water Reactor (MPHWR) undergoing certification.
- APWR (Mitsubishi Heavy Industries)
  Application for certification has been submitted to NRC. Mitsubishi reports that Dallas-based Luminant Power has

selected the US-APWR for its planned new nuclear power plants.

Integral Primary System Reactors (IPSR)

An IPSR reactor is configured to enhance safety of operation by locating the steam generator *inside* the pressure vessel above the core. This reduces by several orders of magnitude the neutron fluence at the pressure vessel compared to current loop-type PWRs. The design objectives include low capital cost, enhanced safety, and proliferation resistance.

- Central Argentina de Elementos Modulares (CAREM)
- CAREM is an Argentine concept to develop, design, and construct a simple, small, and safe nuclear power plant.
- Integrated Modulated Water Reactor (IMR) (Japan Energy Research Institute)

  IMR seeks electricity generation costs comparable to large-scale plants and higher-level safety using passive systems.
- International Reactor Innovative and Secure (IRIS) (Westinghouse BNFL)
- IRIS is in pre-certification process.
- Toshiba 4S (Toshiba)

  This very small, molten sodium-cooled fast reactor with a 30-year metallic alloy core of uranium, plutonium, and zirconium pushes technologies to the leading edge. Toshiba's application to NRC for certification of the 4S is expected in 2009.

Modular High-Temperature Gas-Cooled Reactors (MHTGR)

- Pebble Bed Modular Reactor (PBMR) (Eskom [South Africa])

The PBMR is a small, closed-cycle, gas turbine power conversion system and is regarded as a leader in the power generation field. Its design[184] is presently in "pre-certification" status with the NRC. Application for certification is expected in 2009. Prototype variations of the PBMR design now operate in China and Japan.

- Gas Turbine-Modular Helium Reactor (GT-MHR) (General Atomics in partnership with Russia's Minatom, supported by Fuji [Japan])

  The GT-MHR is a high-efficiency, gas (helium)-cooled reactor moderated with graphite blocks or graphite mixed with fuel in pebbles.

Breeder Reactor

- Liquid Metal Fast Breeder Reactor (LMFBR)

  The LMFBR features a fast neutron spectrum, molten lead or lead-bismuth eutectic (low melting point) coolant, and a closed fuel cycle. A test LMFBR is being constructed in the China Institute of Atomic Energy in Beijing.

Generation IV Reactors (2030 —)

Started by the Generation IV International Forum, Generation IV reactors are generally in research and development (R&D). These reactor designs were undertaken to meet the following objectives:

- Improve nuclear safety;
- Increase proliferation resistance;
- Minimize waste and depletion of natural resources; and
- Decrease construction and operation costs.

With the exception of the Next Generation Nuclear Plant (NGNP), which is scheduled to be completed by 2021, Generation IV reactors are not expected to be ready for commercial use before 2030.

Thermal Reactors

- Very High Temperature Reactor (VHTR)

  The VHTR is a thermal reactor with an expected completion date in or before 2021. Often referred to as the Next Generation Nuclear Plant (NGNP), it is the only Generation IV reactor expected to be on line before 2030. Its high outlet temperature (1,000 ^{0}C) enables applications, such as process heat or hydrogen production.

- Supercritical* Water-cooled Reactor (SCWR)

  The SCWR uses supercritical water, as the working fluid, and has a once-through cycle, like the BWR. It will operate at much higher temperatures than current BWRs and PWRs. The SCWRs are promising, because of their high efficiency, ~45 vs. ~33% and plant simplification.

- Molten Salt Reactor (MSR)

  The MSRs are liquid-fueled reactors capable of burning actinides and producing electricity, hydrogen, and fissile fuels (breeding). First built in the 1950s, there is renewed interest, because of changing goals and technological advances in gas turbines, compact heat exchangers, and carbon-carbon composites.

Fast Reactors

- Gas-Cooled Fast Reactor (GFR)

  The GFR is a fast reactor design, but differs from the HTGR in that it has a fissile fuel content, as well as a non-fissile breeding component, and it has a higher power density than the HTGR.

- Fast Breeder Reactor (FBR)

  The FBR is a fast neutron reactor designed to *breed* fuel by producing more fissile material than it consumes.

- Sodium-Cooled Fast Reactor (SFR)

  The SFR is a project to design an advanced fast neutron reactor for the destruction of high-level waste, plutonium, and other actinides. Building on the LMFBR and Integral Fast Reactor (IFR), SFR is more efficient than thermal reactors with once-through fuel cycles.

- Lead-Cooled Fast Reactor (LFR)

  The LFR is cooled by natural convection with outlet coolant temperatures of 550 – 800 °C. (Lead or lead-bismuth eutectic (LBE) cooling offers enhanced safety and reliability over other liquid metals.) The higher temperature is possible, using advanced materials, and enables the production of hydrogen. Its only drawback is corrosiveness, which results from contact with fuel cladding and certain structural materials.

- Liquid Fluoride Thorium Reactor (LFTR)

  The LFTR is a specific fission energy technology, based on thorium, rather than uranium, as the energy source. The reactor core is in a liquid form at low pressure of ~1 atmosphere, and the reactor is passively safe — it has no control rods. Major advantages include: significant reduction of nuclear waste (it produces no transuranics and achieves nearly 100% fuel burnup); inherent (passive) safety, weapon proliferation resistant, and high power cycle efficiency. It is a derivative of MSR technology, which was successfully demonstrated in the 1960s. The LFTR couples to new closed-cycle turbine power plants with nearly 50% efficiency.

The Generation II BWR and PWR are of interest, only if they are modified to burn reprocessed MOX fuel, or have the potential to burn

MOX fuel derived from weapons-grade plutonium (WPu). Seven advanced reactors that will burn these fuels and contribute to reducing the amount of nuclear waste destined for long-term storage at Yucca Mountain are the following: PBMR, EPR, CANDU-ACR, ESBWR, VHTR, GT-MHR, and LFTR.

PBMR (Westinghouse, PBMR Ltd.)

The Pebble Bed Modular Reactor (PBMR) is a low cost HTGR that uses a direct cycle gas turbine and incorporates passive safety systems. With helium, an inert gas, as its primary coolant, the turbo-machinery does not become radioactive. Its core is based on German HTGR technology and uses approximately two inches in diameter spherical ("pebble") graphite elements containing ceramic-coated fuel particles.[185] The fuel is more highly enriched than the uranium now used in light water reactors (LWRs). South Africa's PBMR (Pty) Ltd, supported by the South African utility Eskom, is the company responsible for development of the PBMR technology. Westinghouse, an investor in PBMR (Pty) Ltd., has taken a leading role in the U.S. design certification; it is presently in a "pre-certification" status with the Nuclear Regulatory Commission (NRC). Because the NRC does not have the same degree of expertise with the PBMR design that it has with LWRs, the process is proceeding somewhat slowly. Application for certification is expected in 2009.

Prototype variations on the PBMR design now operate in China and Japan. Eskom has received approval to build a prototype PBMR in South Africa. If the prototype is successful, Eskom intends to build up to ten local (South Africa) plants and envisions an export market of up to 20 plants a year. With the output about 165 MWe, the PBMR would be one of the smaller reactors now proposed for the commercial market. Some consider this a marketing advantage, because small reactors require lower initial capital investments than larger new units. Of significance as well, the smaller units are less attractive as targets for terrorists. Several PBMRs could be built at a single site, as local power demand requires, which would also be very cost-effective since one control room and one set of maintenance workers could be utilized.

EPR (AREVA NP)

The European Pressurized Water Reactor, or *Evolutionary* Pressurized Water Reactor (USEPR), as the U.S. design is known, is a large PWR of evolutionary design, using proven technologies. It has four separate, redundant (or 100% capacity trains) safety systems, double-walled containment, and a "core catcher" for containment and cooling of core materials in the event of reactor vessel failure. Its design relies on active safety systems, not on passive safety features. Average availability (reliability) is expected to be 92%. Expected economic impacts include a 10% reduction in operating costs and 17% less uranium per kWh over current LWRs. The first EPR is under construction in Finland with a target date for commercialization during 2010. The second is in France, where the government has authorized building an EPR at Flamanville 3. Construction is scheduled to be completed in 2012. EPRs might replace additional commercial reactors now operating in France, starting in the late 2010s, and AREVA NP (Framatome) hopes to build in China and elsewhere, as well.

In the United States, UniStar Nuclear Energy, LLC, a new holding company comprising Constellation Energy, Electricite de France, and Bechtel, is the leading nuclear fleet owner and operator in North America. The U.S. EPR (USEPR) is UniStar's technology of choice, because it sets new standards for safety, efficiency, and performance. Constellation Energy has the right to develop nuclear projects at its Calvert Cliffs Nuclear Plant in southern Maryland, and at the Nine

Mile Point Nuclear Station and R.E. Ginna Nuclear Plant, both of which are located in upstate New York.[186]

The proposed size for the EPR has varied over time, but is most frequently placed around 1,600 MWe. Earlier designs were as large as 1,750 MWe.

CANDU-ACR (Atomic Energy of Canada Limited [AECL])

Also known as the Advanced CANDU Reactor, ACR700 or ACR1000, the CANDU-ACR is a Modified Pressurized Heavy Water Reactor (MPHWR).

The AECL's ACR series of reactors is considered by its vendor to be an evolution from the internationally successful CANDU line of PHWRs. Original pre-application design certification procedures in the United States had been for the 700 MW ACR700 design. More recent discussions have focused on the 1200 MW ACR1000. The CANDU reactors and their Indian derivatives have had more success than any family of commercial power reactors except the LWRs. One of the innovations in the ACR series of reactors, compared to earlier CANDU designs, is that heavy water is used only as a moderator in the reactor. Light water is used as the coolant. Earlier CANDU designs used heavy water both as a moderator and as a coolant.

Fueling procedures for the ACR follow the earlier CANDU designs in that it occurs while the reactors are in service rather than during refueling outages. The AECL has aggressively marketed the ACR series offering low prices, short construction periods, and favorable financial terms. As is the case for most non-LWR reactors, U.S. generating companies, nuclear engineers, and regulators have limited familiarity with the design. Early interest in the ACR series by Dominion Resources in Virginia and by United Kingdom generating companies has not been sustained. Subsequently, AECL delayed its efforts to certify the design in the United States. The CANDU-ACR is a contender among designs, being considered for construction in Ontario. The earliest possible reactor construction there might be either the earlier CANDU designs or non-Canadian designs.

ESBWR (General Electric)

The Economic Simplified Boiling Water Reactor (ESBWR) is a new, simplified BWR design promoted by General Electric and some allied firms. The reactor constitutes an evolution and merging of several earlier designs, including the ABWR. The ESBWR, which includes new passive safety features, is intended to cut construction and operating costs significantly from earlier ABWR designs. Its approximate electrical capacity is 1,550 MWe.

General Electric and others have invested heavily in the ESBWR, and two U.S. utility companies, Entergy and Dominion, plan to use the reactor in their new construction plans. Entergy, an integrated energy company engaged in generation and distribution of electricity, is the second largest nuclear generator in the United States with ten nuclear power plants in seven states. Entergy plans new construction nuclear plants at its Grand Gulf, Mississippi, and River Bend, Louisiana sites. Dominion Energy, which serves retail customers in 11 states in the Midwest, Mid-Atlantic, and Northeast. plans to construct a new reactor at its North Anna site. These utilities have applied for a combined construction and operating license (COL) for new ESBWR reactors. As of 30 June 2008, their license applications were "under review."[187]

VHTR (General Atomics et al.*)

The Very High Temperature Reactor (VHTR) will be an advanced, very high temperature (approximately 1,000 °C coolant outlet temperature, which is beyond capabilities of conventional reactors), graphite moderated, helium-cooled reactor with a once-through uranium fuel cycle. Based on demonstrated technology (Fort St. Vrain and Peach Bottom prismatic reactors and German pebble bed), the VHTR is the nearest term of six Generation IV reactor technologies for nuclear assisted hydrogen production. Environmentally friendly, the VHTR has no carbon emissions, and it has the flexibility to adopt uranium and plutonium fuel cycles.

Another and perhaps primary use of VHTR or MHTGR is as a heat source. The Canadians and Dow Chemical are looking seriously at this application. Dow Chemical's utility bill is $5 billion a year, and the company is concerned that a carbon tax may be levied in the future. Also, Canadians are interested in this for use in oil sands projects. For Dow Chemical, it's already known that Entergy would run it and sell the processed heat to Dow Chemical.

The DOE's primary mission for the VHTR is to demonstrate nuclear reactor assisted cogeneration of electricity and hydrogen while meeting the Generation IV goals for safety, sustainability, proliferation resistance and physical security and economics. Successful deployment of the VHTR as a demonstration project will aid in restarting the atrophied U.S. nuclear power industry. Project participants include DOE laboratories, industry partners, designers, constructors, manufacturers, utilities and Generation IV international countries. The VHTR Project includes an overall reactor design and construction activity and four major supporting activities: fuel development and qualification, materials selection and qualification, NRC licensing and regulatory support, and the hydrogen production plant.

(*Note: General Atomics is partnered with U.S. National Laboratories: Los Alamos, Idaho, Oak Ridge, along with TransWare, Inc. [California], and Studsvik-ScandPower AB [Sweden]).

GT-MHR (General Atomics)

The Gas Turbine Modular Helium Reactor (GT-MHR)[188] and the Freedom Reactor (Entergy trademark) are relatively small, high temperature gas-cooled reactors (HTGRs) that have an approximate capacity (electric) of 285 MWe. Entergy is a supporting generating company for development only. The most advanced plans for GT-MHR development relate to building reactors in Russia to assist in the disposal of surplus plutonium.[189] Parallel plans for commercial power reactors would use uranium-based fuels enriched to as high as 19.9% U_{235} content. This would keep the fuel a fraction below the 20% U_{235} enrichment that defines highly enriched uranium.

A research version of the reactor will be constructed for the University of Texas Permian Basin and affiliated institutions at Andrews County, Texas. Because coolant temperatures arising from HTGRs are much higher than from LWRs, the design is viewed as a potential source of commercial heat. Particular attention has been paid to the design's potential to produce hydrogen from water. The GT-MHR is considered, among many other designs, as a potential contender for the U.S. Department of Energy's Next Generation Nuclear Plant (NGNP) program. The NRC design certification status is Pre-application Review.

LFTR (ORNL R&D)

The LFTR technology is derived from the Molten-Salt Reactor Program (MSRP), which began at ORNL in 1958. The MSRP was permitted by the AEC to build a small reactor provided it produced less than 10 MW of thermal power. Design and construction of the Molten-Salt Reactor Experiment (MSRE) was a "true" liquid-fluoride power reactor begun in 1961. It utilized a lithium7-beryllium fluoride solvent that dissolved zirconium and uranium tetrafluorides. The MSRE was designed to simulate the "core" of that future "thorium breeding" reactor, but that objective was deferred, because the favored design was a two-region liquid-fluoride breeder.

Advantages of the LFTR include the following:

- Inherent safety. The strong negative temperature coefficient of the fluid fuel, its response to transients, the stability of fission products in the salt, and the ability to drain the core into a passively-cooled configuration led many to consider the LFTR the *safest reactor ever designed*. But, the passive safety issue was not a primary concern in the 1970s when the LFTR was compared to the liquid metal fast breeder reactor (LMFBR). Further, passive safety usually results in drastic reductions in reactor performance.

- High performance. The LFTR can operate at the high-temperatures and low pressures needed for high-efficiency electrical production and thermochemical hydrogen production.

- Operability and reliability. Its capability to be refueled continuously while online eliminates LFTR refueling shutdowns thus improves the competitiveness of commercial utilities — a big plus.

- Security. Properly-designed, the LFTR can withstand severe accidents, such as a breach of vessel and containment, whether intentional or accidental. If the fuel salt were inadvertently exposed through a combined breach of containment and vessel, the salt would freeze and occlude fission products in the salt as stable fluorides.

Having examined the advanced technology reactors being developed by U.S. companies and laboratories that can be available in the near- to mid-term, the GT-MHR appeared to offer the best answer toward resolution of the nuclear waste disposal issue. Specifically, it is small, modular, and passively safe;[190] it can burn reprocessed MOX and weapons-grade plutonium; it can be factory produced in quantity and sited underground; and, it is resistant to air attack and terrorist invasion — for these reasons, the GT-MHR will be discussed in more detail in following paragraphs.

The GT-MHR design was possible because of four recent technical advances:

- Modular helium reactors with inherent safety characteristics are capable of producing very high gas temperatures;
- High-efficiency gas turbines developed for the airline and utility industries;
- Plate-fin heat exchanger technology; and
- Frictionless magnetic bearings

The result is a simplified power cycle with very high efficiency and reliability, and low power cost.

The inherently passively-safe GT-MHR power plant includes one or more modular units in underground silos, each containing a reactor vessel and a power production vessel. It works because helium is naturally inert and single phased. It can operate at higher temperatures than today's conventional nuclear plants. The higher the turbine's operating temperature, the more efficient the plant becomes, and the less heat released to the environment.[191] Helium directly drives the turbine, instead of going through a large heat exchanger to produce steam. The combination of the MHR and the gas turbine represents simplicity, safety, and economy. The reactor coolant directly drives the turbine that turns the generator. This eliminates costly and failure prone steam generation equipment. There is no corrosion that causes leaks, reduction in operating life, or stress corrosion that causes structural failures.

T.A. Donohue (General Electric) once observed that the Gas Turbine-Modular High Temperature Reactor turbo machinery is a logical application of GE's successful jet engine and power turbine technology.[192] Engine and reactor sizes are similar, and stresses, temperatures, and pressure are either less demanding or comparable. Because it is inert, helium is an excellent working fluid that eliminates oxidation and corrosion. Its properties provide subsonic flow fields throughout the machine and eliminate the complexities of transonic and supersonic flows in the blading. The GT-MHR magnetic bearings are a modest extension of existing in-service technology. They are essentially frictionless and provide automatic and adjustable dynamic damping and on-line monitoring resulting in improved performance and reliability. Of particular importance are the elimination of oil-lubricated bearings and the potential ingress of oil into the working fluid. All things considered, the GT-MHR appears to be a rational, practicable, and economic approach to the next generation of nuclear power plants.

Dr. J. A. Fridericy of Allied Signal Aerospace stated, "The recuperators (same as condensers in LWRs) for the GT-MHR are about the same size as units we have made for the fossil fuel power industry. In fact, we have made some 2½ million units using this type of construction. At least 60 have been for large gas turbine plants. These units utilize approximately 1,000 individual brazed modules. GT-MHR temperatures are less demanding than units now in operation, and efficiencies are within the range of units previously delivered. Pressures are higher, but that is not viewed as a problem. The non-corrosive helium environment is very beneficial."[193]

The entire GT-MHR power plant is essentially contained in two interconnected pressure vessels enclosed within a belowground, concrete containment structure. The larger vessel contains the reactor system and is based on the MHR that was designed as part of the DOE Modular High-Temperature Gas-Cooled Reactor (MHTGR) program. The second smaller vessel contains the entire power conversion system. The turbo-machine consists of a generator, turbine, and two compressor sections mounted on a single shaft rotating on magnetic bearings. These bearings control shaft stability while eliminating the need for lubricants within the primary system. The vessel also contains three compact heat exchangers. The most important of these is a 95% effective plate-fin recuperator that recovers turbine exhaust heat and boosts the plant's efficiency from 34% to 48%. As an additional benefit, the GT-MHR has the potential to burn fuel derived from weapons-grade plutonium (WPU) as fuel to provide electrical energy.

The GT-MHR was the first reactor design that met the criteria of the International Nuclear Event Scale (INES) Level 1 safety standard. This is a scale developed jointly by the Nuclear Energy Agency (NEA) and the International Atomic Energy Agency (IAEA) to classify events at nuclear power plants. Level 1 is the lowest level; any lesser event is "below scale." Chernobyl is classified Level 7, Major Accident; Three Mile Island Level 5, Accident With Off-Site Risk. At TMI, the off-site release of radioactivity was very limited. The event is classified as Level 5, based on the on-site impact.[194]

The GT-MHR produces electricity at lower costs than the LWRs and is competitive with next generation "clean" coal-fired plants. The DOE calculated in 1986 that power from such coal generators would cost an average of 5.5 cents per kWh. Power from modular reactors can be brought to market for 4.5 cents per kWh. These savings can be realized, because the GT-MHR will be factory produced to a single, pre-licensed design. Construction costs are expected to be less than $1,000 per kilowatt, whereas Seabrook (New Hampshire) and Shoreham (Long Island) were more like $5,000 to $6,000. Also, the operation and maintenance (O&M) costs of LWR plants are much higher, because they require very large staffs — 700 people per plant vice the 350 or so that will be required to staff the GT-MHR. (This savings is based on a plant containing four or more standardized reactor modules at the same location.) Galushkin, stated, "The cost [in U.S. dollars] of the GT-MHR 4-module commercial nuclear power plant (NPP) is less than $900 per kilowatt electric (kWe)."[195]

The GT-MHR will not only be much cheaper to build, but the high efficiency means there will be less heat thrown away.[196] However, Generation IV reactors will not eliminate the waste disposal problem, but their ceramic encapsulated fuel does simplify it. The fuel, which can survive unscathed in a reactor core during an accident, is securely packaged for storage conditions that are essentially benign. Many problems associated with LWRs are eliminated. The amount of nuclear waste destined for Yucca Mountain is reduced significantly because of the reactors higher efficiency.

For example, the GT-MHR is a new turbine generating power system powered by a passively, safe nuclear reactor. It eliminates the need to make steam to produce electricity and, like all nuclear reactors, eliminates the pollution and waste of fossil-fuel generating plants. It could help to reduce our billion dollars a week deficit for foreign oil as well. The result is a simplified power cycle with very high efficiency and reliability and low power cost. Thermal efficiencies are increased by 50% relative to conventional nuclear reactor plants. Current reactors produce 50% more high-level waste per kWh of electricity than the GT-MHR.

The GT-MHR is designed to have multiple layers of tough, high temperature tolerant pyrolytic carbon and silicon carbide that confine the radioactive fission products at their source — in the center of the fuel particle. It consists of pyrolytic carbon, silicon carbide, porous carbon buffer, and uranium oxycarbide. These coated fuel particles, identified as TRISO-coated particle fuel, are formed into fuel rods and inserted into graphite fuel elements. Tests in the United States, Europe, and Japan have demonstrated TRISO-coated particle fuel to have high potential for retention of fission products to temperatures approaching 2,000 °C, well above the 1,600 °C worst-case accident temperature of the MHR.

Like other U.S. power reactors, the GT-MHR has a negative temperature coefficient. By contrast, Chernobyl had a positive reactivity coefficient. Its temperature increase acted to intensify the fission reaction, resulting in a runaway. The heat results from the decay of fission products. The heat does not stop when the power is shut off. So, having a negative temperature coefficient is good, but it is not good that fission reaction has stopped because of the loss of the cooling water. With the GT-MHR, its decay heat will not cause a meltdown, even if the coolant is lost. The reactor's low power density and geometry assure that decay heat will be dissipated passively by conduction and radiation without ever reaching a temperature that can threaten the integrity of the ceramic-coated fuel particles, even under the most severe accident conditions.

Because the GT-MHR is passively safe, it does not require active safety systems characteristic of current LWRs. The gas turbine eliminates the complex, hard-to-maintain steam generators common both to nuclear plants and ordinary fossil-fired power plants. The GT-MHR is a power plant that produces electricity at lower costs than other nuclear reactors. (It was supposed to be competitive with the cost of the next-generation or "clean" coal-fired plants, but funding for the clean coal technology is being pulled and interest in clean coal technology is rapidly going away. The DOE calculated in 1986 that power from such coal generators would cost an average of 5.5 cents per kilowatt-hour, much higher than nuclear.)

Conventional, low-temperature nuclear plants operate at 32% thermal efficiency. The GT-MHR power plants can achieve thermal efficiencies of approximately 50%, perhaps even higher in the future. As a result, there is much less thermal discharge into the environment. There is a dramatically lower high-level radioactive waste per unit of energy. Today's reactors produce 50% more high-level waste than will the GT-MHR. The GT-MHR plants can use air-cooling or water-cooling.

To summarize the case for the GT-MHR, the following are pertinent:

- The GT-MHR design is derived from natural properties of materials and optimum choice of reactor size, geometry, and power density.
- It can withstand the total loss of coolant without the possibility of a meltdown; the operating safety risk is virtually zero.[197]
- The GT-MHR could be a major source of environmentally sound energy for the United States.
- The core is meltdown-proof, and it provides safety through the laws of nature. It is inherently safe.
- As with all nuclear power plants, no carbon dioxide, no acid rain, a hundred thousand times less waste volume than coal, and it preserves natural resources.
- The GT-MHR provides an answer, at least in part, to the pressing national issue of nuclear waste disposal.

Shortly after 9/11, when research for this book was undertaken in earnest, it was anticipated that the GT-MHR would be operational by 2010, but the program has not proceeded as envisioned, and its probable deployment date is at least 15 years into the future, perhaps even 2030. Even if DOE accelerated the program, it would be too late to have a near-term effect on the resurgence of nuclear power. But its design characteristics are exceptional, particularly well-suited to meet many of the challenges facing our nation today.

Meanwhile, interest in the thorium-fueled LFTR, piqued by Dr. Teller's final paper (co-authored with Dr. Moir) and given impetus by Thorium Power's development of thorium, uranium, and plutonium mixed fuel assemblies (being tested in Russia this year), give reason to consider the possibility that the LFTR may well be a viable competitor to the GT-MHR.

The two reactors share most of the desired characteristics necessary to achieve the objectives proposed, and it appears likely that both will be available in the same time frame. Because of their small size and simple design, the GT-MHR and LFTR reactors are ideally suited for modular construction, off-site in factories. The economies of scale are replaced by economies of serial production. Significant benefits of factory production include the following:

- Reduced construction time
  - Years to months;

- Reduced interest carrying costs
  - Years to months;

- Increased efficiency of labor; and
- Use of robots on reactor assembly lines.

These two advanced reactors, the Generation III⁺ GT-MHR* and Generation IV LFTR, are especially interesting because they can be factory produced and sited underground. (*In earlier copyrighted versions of this book, the GT-MHR was classified Generation IV; it currently appears in most listings as Generation III⁺) Both the GT-

MHR and the LFTR designs incorporate tested technologies; no breakthrough innovations in technology are required. Unlike their LWR predecessors, these reactors are more efficient, inherently safe, and far less vulnerable to attack by terrorists.

A note on reactor safety is in order before leaving this discussion. The situation to date, August 2008, is that in over 12,700 reactor years of civil operation in 32 countries, there has been only one major accident to a commercial reactor, which was not contained within the design and structure of the reactor. The Chernobyl (Ukraine 1986) power runaway was initiated by human error that resulted in loss of coolant, steam explosion, and intensive fire that led to structural failure. The accident killed 31 people at the time; the toll to date is 56, possibly more. There were health and environmental consequences as well. A second accident, TMI (Pennsylvania 1979), also caused by human error, resulted in loss of coolant and partial core meltdown, but radioactivity released from the reactor vessel was confined within the containment. There were no health or environmental consequences. In its 57-year history, nuclear power has caused fewer deaths than any other major energy power source.

Despite this record, the public's attention is immediately driven toward the subject of safety when nuclear power is mentioned. Largely as a consequence of TMI, the public remains skeptical even though there were no injuries at or in the vicinity of TMI. In addition to *safety*, however, particularly as a consequence of 9-11, one must also consider *security* and *safeguards* and their interrelationships. In that context, consider the following:

- Safety relates to intrinsic hazards, such as accidental core meltdown and/or release of radioactive particles or rays;
- Security relates to external threats; intentional theft and misuse of nuclear materials; and

- Safeguards are actions taken to preclude activities by terrorists or rogue nations that could lead to acquisition of nuclear weapons or fabrication of "dirty bombs."

Looking to the longer-term future, the newest development in power to meet our gluttonous energy demands, is the International Thermonuclear Experimental Reactor (ITER) fusion energy project. The ITER is a joint venture of China, the European Union, India, Japan, the Republic of Korea, the Russian Federation, and the United States, which was officially implemented by agreement signed in Paris in November 2006. Its mission is to demonstrate the technical and scientific feasibility of fusion power. Notably, the first stand-alone, large-scale, international scientific research project in history is to be constructed in Cadarache, in the South of France.

CHAPTER 8

Spent Fuel, Recycling, Reprocessing, and Vitrification

Seize the day, put no trust in tomorrow.
Horace[198]

Spent Fuel

In a nuclear reactor, producing heat and then electricity involves fuel assemblies that are precisely geometrically arranged that make up the core of the reactor. For example, a 900 megawatts electric (MWe) pressurized water reactor core contains 72.7 tons of enriched uranium. The fuel is progressively consumed and transformed, and it must be replaced regularly. Each fuel assembly stays in the reactor core for three or four years (a quarter to a third of the assemblies are replaced every 12 to 18 months). During the operating period, fission of the uranium provides the heat required to produce electricity. The fuel is changed, which makes it gradually less efficient in the following ways:

- Uranium-235 content decreases through fission;

- Plutonium, formed by capturing neutrons from U_{238} atoms, produces energy just like uranium; and
- Fission products and actinides are formed. These make up the final waste from the reaction since they cannot be recycled like the other products. Some fission products make the fuel less reactive and are given the name of poison isotopes; however, *not all fission products are "poison."*

When the fuel is too spent to maintain a nuclear reaction and to produce energy in an efficient manner, it must be removed from the reactor core and replaced with fresh fuel. However, the spent fuel still contains a large proportion of recyclable, energy-rich materials.

The fission products make the radioactivity of the spent fuel higher than that for fresh fuel and require special handling precautions. Prior to recovering materials from the spent fuel, the assemblies removed from the core remain for at least one year in a pool next to the reactor, where the fission products become less radioactive. Water provides an effective shielding against radiation and cools the assemblies, which continue to release heat for a time. At the end of this temporary storage when activity has decreased, the spent assemblies are transported to the reprocessing plant. For example: one light water fuel assembly consists of 500 kilograms (kg) of uranium prior to irradiation in the reactor. After irradiation, the percentage depends upon burn-up level. There are 475 to 480 kgs of uranium, and 95 to 96% can be recycled. After irradiation, there are 5 kgs of plutonium or 1% that can be recycled. (These are recyclable materials.) In addition, there are 15 to 20 kgs of non-recyclable fission products with 3 to 5% terminal residues considered waste.

Spent fuel contains 97% recyclable materials. Whereas, fresh fuel contains 3 to 5% U_{235} and 95 to 97% U_{238}, the same fuel after use in the reactor core still contains 1% U_{235}, 95% U_{238}, 1% Pu, and about

3% non-recyclable wastes. Therefore, spent fuel contains nearly 97% materials that can be recovered and reused upon reprocessing and recycling.

Clearly, recycling and reprocessing offer major advantages — savings in raw materials and reductions in the volume and toxicity of wastes. Management of nuclear wastes must be integrated from the beginning, and the process constantly monitored and improved. Uranium in spent fuel has properties comparable to mined uranium concentrates. Its recycling saves the equivalent amount of natural uranium. Plutonium has a high-energy value in its first recycling: *1 gram of plutonium = 100 grams of uranium = 1 ton of oil.* Recycling doubles natural uranium savings while destroying plutonium. In addition, *5,000 tons of recycled spent fuel produces the same amount of electricity as the combustion of 100 million tons of oil — the total annual output of Kuwait.*

Reprocessing recovers plutonium that is fissile material. Recycling in the manufacture of mixed oxide (MOX) fuels uses its great energy potential. Uranium, also recovered by reprocessing, still contains about 1% of U_{235}. Stored in a stable form, it can be enriched and again recycled.

The object of plutonium separation is its recycling into MOX fuel. The plutonium separation process produces a constant quality plutonium oxide-2 for homogeneity, specific area, particle size, sintering ability, and moisture content. The manufacturing of MOX fuel can be performed with a well-characterized material. (Production of MOX fuel requires that the chemicals in the plutonium and uranium oxides be identified precisely, thus the term "characterized.") In MOX fuel, this fissile product, plutonium, takes the place of U_{235} in a fresh fuel. Plutonium, which is a fissile material, is part of the composition of MOX fuels (mixed oxides). The isotopic composition of plutonium recovered from pressurized water reactors and boiling water reactors spent fuel, along with the buildup of americium, makes the waste radioactive. In reprocessing plants, the equipment is housed in shielded hot cells, and the process is remotely operated. Maintenance work is performed only after the plutonium has been flushed out. Because the equipment requiring maintenance is located out of personnel accessible areas, most maintenance tasks can be performed without flushing out plutonium.

As mentioned earlier (Chapter 1), by 2007 the United States had generated 70,000 metric tons of heavy metal (MTHM) of spent nuclear fuel. When recycling spent nuclear fuel, the resulting final high and low level radioactive residues or the non-reusable part of the fuel (fission products and concrete-encapsulated or compacted hulls and end-fittings) is packaged or vitrified. Residues (waste packaged for final storage) are processed, according to very stringent technical specifications approved by international regulatory bodies.

France's La Hague reprocessing plants, UP_2 (renamed UP-800 in 1994) and UP_3, utilize the most advanced technologies to achieve six goals:

1. High industrial performance (availability and operating cost);
2. Maximum recovery of uranium and plutonium;
3. Low radiological exposure to operating and maintenance teams;
4. Low environmental impact;
5. Advanced waste solidification (low volume and low toxicity); and
6. Advanced safety requirements.

The metric tons of uranium (MTU) recycled as of November 1, 1999, are shown by country per Table 1.

Table 1 **Metric Tons of Uranium Recycled**

	MTU* At UP$_2$ Plant	MTU At UP$_3$ Plant	Total MTU
France	6,541		6,541
Germany	1,643	2,477	4,120
Japan	151	2,690	2,841
Switzerland	132	430	562
Netherlands	85	141	226
Belgium	139	453	592
U.S., Sweden, and Finland	0	0	0
TOTAL	8,691	6,191	14,882

*Metric tons(s) uranium

Plutonium oxide (PuO_2) is recycled through a MOX fuel fabrication plant where it is mixed with depleted UO_2 to make fresh fuel. The high-energy value of the plutonium is thus recovered when the MOX is used to fuel LWRs.[199] France's La Hague Plants achieve approximately 99% separation efficiency for uranium and plutonium. [200]

Radiation protection (safety of employees) program statistical data dating from 1976 reflect the success of the French nuclear operation of its UP_2 and UP_3 plants. In 1998 the average annual personnel exposure (including operation and maintenance) was 0.55 mSv*, while natural radiation exposure is much larger. A nuclear worker is permitted between 20 and 50 mSv per year. (* Note: "mSv" is a measure of radiation exposure.)

Operational experience with MOX fuel is growing rapidly throughout Europe, Asia, and the Far East. Over 30 thermal reactors in Europe are using MOX, and an additional 20 have been licensed. In France, *Electricite de France* (EDF) plans to have all of its 900 MWe series reactors running with at least 30% MOX. Japan plans to have one-third of its current reactors using MOX by 2010, and a new reactor capable of a complete fuel loading of MOX is under construction.

The operating nuclear reactors in Europe are shown by type in Table 2.

Table 2 **Reactors in Europe**

	Reactors in Operation	MOX Reactors*	"Moxified" Reactors*	First MOX Loading Date
France	57	20	18	1987
Germany	21	11	10	1972
Belgium	7	2	2	1995
Switzerland	5	4	3	1984

*MOX reactors that use mixed oxide fuel are called "moxified reactors." Originally designed to use enriched uranium fuel, the reactors have been modified to use mixed oxide fuel.

MOX fuel is approaching enriched uranium fuel in terms of performance in the following areas:

- High burn-up (52,000 MWd/MTHM);
- Quarter-core reloads management; and
- Load-following and grid-following.

MOX performances (measured in MWd/t) follow UO_2 fuel performances with a few years' difference. They are as follows:

Table 3 **MOX Fuel Performances**

	UO_2 1985	UO_2 1995	UO_2 1999	MOX Series	MOX Prototype
Burn-up*	35,000	47,000	52,000	39,000	46,500**
Load	Yes	Yes	Yes	Yes	Yes
BurnupTrend	Upward	Upward	Upward	Upward	Upward

*MWd/MTHM
**The next cycle is 51,400 MWd/MTHM

It is very important to note that plutonium recycling with MOX fuel controls plutonium inventories because it consumes Pu, reduces new production, and generates kWh at a constant or even declining worldwide plutonium balance. The more reactors are "moxified," the lower the annual plutonium production. Recycling with MOX fuel reduces waste volume and toxicity from the first recycling. Spent fuel elements are reduced by a factor of 8, and multiple recyclings reduce plutonium contained in ultimate waste by 100. Plutonium recycling with MOX fuel also conserves large amounts of natural uranium and oil.

Table 4 **Plutonium Balances in Large PWRs**

The comparison of annual balances of plutonium for PWR reactors producing 50,000 MWe is as follows:

	100% UO$_2$ Reactor	30% MOX/70% UO$_2$ Reactor	
Input Pu	0 kg	350 kg	
Output Pu	200 kg	350 kg	
Final Balance Pu	200 kg	0 kg	
	%UO$_2$	%MOX	Final Pu
or	100 %	0%	10 t/yr
50 Reactors of	60%	40%	6 t/yr
1,000 MWe	0%	100%	0 t/yr

Based on these data, the conclusion can be made that the more reactors that are loaded with MOX, the smaller the amount of plutonium generated (annually). It takes about 10 years to complete the fuel cycle back-end process. From a non-proliferation standpoint, MOX fuel is safer than enriched uranium fuel, both un-irradiated and irradiated.

Recycling

Recycling enables a nation to recover the usable plutonium and uranium in spent fuel to the equivalent amount of fresh plutonium and uranium. The power plant operator can concentrate on its core business, which is electricity generation. Since France began reprocessing in 1966, effluent and waste volumes have been continuously reduced. Today, recycling a ton of spent fuel generates less than $0.5m^3$ of high or medium level waste.[201] The process is shown in Figure 2.

To preserve resources, reduce nuclear waste and the amount of surplus plutonium, spent fuel should be recycled to recover the unused uranium-235 and plutonium produced in nuclear reactors. It should also be noted that some plutonium byproducts formed during reactor operations, such as plutonium-240 reduce the effectiveness of plutonium as a bomb material by forcing more specialized separation of the Pu_{239} isotope. Recycling is a service industry. It is important to note that throughout operations, nuclear materials contained in spent fuel remain the property of the client power companies. Therefore, reprocessing contracts specifically require that conditioned final residues be returned to the country of origin.

Reprocessing

Reprocessing is the separation of spent fuel components for recycle or storage of non-reusable wastes using maximum safety and small volumes. France and other counties, where nuclear energy is a priority, chose reprocessing as an efficient method to manage the back-end of the nuclear fuel cycle. It offers three advantages:

1. Recovers 96 to 97% of the energy-rich materials still contained in the spent fuel (U and Pu);

2. Facilitates temporary storage or terminal disposal of non-reusable wastes that are isolated to reduce their volume and then suitably conditioned; and

3. Reduces the toxicity of terminal wastes since the plutonium that is highly radiotoxic is reused rather than considered a waste.

As a result, reprocessing contributes to the long-term protection of the environment.

The plutonium present in the world today is all man made. It is formed by the reaction of uranium fuel in nuclear reactors. During this reaction, U_{238} captures a neutron and transforms into U_{239}. In turn, U_{239} transmutes into neptunium-239 (Np_{239}). Then, every other day, half of the Np_{239} transforms into Pu_{239}.

In traditional reprocessing, spent fuel is dissolved in acid, and then the uranium, plutonium, and other fission products are separated from the mixture. The uranium can be re-enriched and recycled. The plutonium is recombined with U_{238}, made into rods, and put into reactors. The fuel is called "mixed oxide," or MOX, and essentially substitutes Pu_{239} for the fissile U_{235} in first generation fuel. The other fission products are encased in glass and stored.

Figure 2 Schema:
the Recycling Process (Fuel Fabrication)

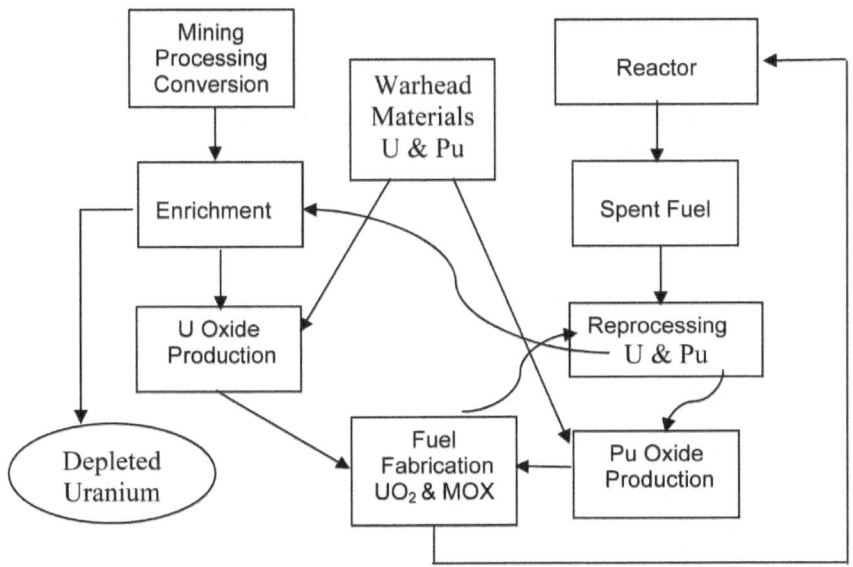

The MOX process can be used to dispose of weapons-grade plutonium "pits" originally intended for use in atomic warheads. The process converts the pits into powder, then blends powder with uranium oxide into fuel pellets that are loaded into rods that are burned in commercial reactors.

In recycling the final high-level waste (HLW) and low-level waste (LLW) residues, the non-reusable parts of the fuel (vitrified fission products and concrete-encapsulated or compacted hulls and end-fittings [structural waste, such as concrete {90 –95}]) are packaged under internationally agreed specifications. The Nuclear Regulatory Commission (NRC) approval of specifications came August 1, 1996. These specifications have been approved by France, Japan, Germany, Belgium, Switzerland, and Netherlands. The waste of the hulls and end-pieces were approved for France, Japan, Germany, Belgium, Switzerland, and Netherlands. The method and machine compacting of waste material, in particular metallic scrap, is under progress. Residues (waste packaged for final storage) are processed as genuine industrial products, according to very demanding technical specifications approved by international regulatory bodies. These are placed in Universal Canister (UC-C) (glass canisters), and the hulls and end-fittings are placed in technological Universal Canisters.[202]

Based on Le Marche du combustible nucleaire-AREVA-DSI-Updating 1996, the spent fuel management situation in countries with large nuclear programs shows spent fuel (**figures in metric tons based on oxides) is located:

Table 5 **Reprocessing Statistics**

	1996-1997	1998-2000	2001-2005
WOCA* nations' reprocessing capacity * World Outside the Centrally-Planned Economies Area	2,590**	2,590	3,390
Spent-fuel discharges in Europe and Asia, except Sweden	3,604	3,672	3,773
Spent-fuel discharges in USA and Sweden	2,389	2,262	2,163

The uranium and plutonium separation efficiencies in France's La Hague Plants are as follows:

Table 6 **Separation Efficiencies (Uranium and Plutonium)**

	Uranium	Plutonium
Uranium and plutonium recovered	99.88%	99.88%
Glasses and other wastes	0.12%	0.12%
TOTAL	100%	100%
The by-products are as follows (Except Tritium)		
Vitrified waste	99.5%	97.6%
Hulls and end-fittings	0.4%	2.3%
TOTAL PROCESS WASTE	99.9%	99.9%
Technological waste	0.1%	0.1%
TOTAL	100%	100%

Table 7 **Back-End Costs**

The cost (in French centimes[c] per kWh, discount rate 0%) of the fuel cycle back end was as follows:

	Closed cycle	Open cycle
Spent fuel transportation	0.10	0.10
Spent fuel storage		0.31
Reprocessing and vitrification	1.20	
Spent fuel packaging		0.69
Waste disposal	0.11	
SUBTOTAL		
Credits:		
Uranium credit	-0.18	
Plutonium credit	-0.07	
SUBTOTAL	-0.25	
TOTAL COST	1.16	1.10

Table 8 **Strategic Choice**

The fuel cycle back-end cost for 50,000 MWe reactor fleet is as follows:

	Minimize plutonium quantities in final residues	Accept significant plutonium quantities in final residue
Closed cycle	1.16 c/kWh* (10 kg/y of Pu)	0.8 c/kWh** (100-200 kg/y of Pu)
Open cycle	Forbidden	1.10 c/kWh (10,000 kg/y of Pu)

*Information from OECD 1994

**Information was extrapolated from current reprocessing prices.

Both of these options seem to cost the same; however, closed fuel cycle costs are based on FIRM SALES PRICES for reprocessing services while open fuel cycle costs are ESTIMATED PRODUCTION COSTS for underground storage facilities that will not operate for some 20 years. These options are not comparable in terms of environmental impacts. In reprocessing, the plutonium amount in the waste is reduced to comply with the ALARA principle. (The International Commission on Radiological Protection [ICRP] recommended system for limiting the doses received by persons.)

The MOX fabrication facilities recycle plutonium recovered through reprocessing. AREVA supports operation of three MOX fuel fabrication plants: The Melox plant in Marcoule, France (start-up 1995); the AREVA NC plant in Cadarache, France; and the Belgonucleaire plant in Dessel, Belgium. There are 35 reactors in Europe that have been loaded and successfully operated with MOX fuel. Europe embraces recycling. The MOX fuel has been used for many years in light water reactors, since:

- 1982 in Germany;
- 1985 in Switzerland; and
- 1987 in France.

The capacities of AREVA's MOX and UO_2 fuel fabrication plants in tonnes of heavy metal (MTHM), uranium + Pu, per year (tHM/y) are demonstrated in Table 9:

Table 9 AREVA Fuel Fabrication Capacities

Country	Company	Location	Process Spent Fuel	Fabricate and Market	
				MOX	UO$_2$
France	AREVA NC	La Hague	1,700		
France	Melox	Marcoule		195	
France	AREVA NP	Romans			150
France	AREVA NC	Cadarache		40	
Belgium	Belgonucleaire*	Dessel		40	

*Production marketed by AREVA NC

Note: In this discussion, metric tonnes are given: equivalent to ~2,205 pounds.

The AREVA MOX fabrication capabilities involve the facility in La Hague that can process 1,700 tonnes per year spent fuel, and Marcoule that can fabricate 195 tonnes per year of MOX fuel. Reprocessing at La Hague of spent fuel from nuclear power plants in France and other countries extracts 99.9% of the plutonium and uranium for recycling. The remaining 3% of the spent fuel material is high-level toxic waste that must be vitrified and stored. Recycled uranium (RepU) can be enriched and converted at the Romans plant into UO_2 fuel. The EdF has demonstrated its use in two of its reactors and increased utilization is a feasible option. It sends 850 tonnes per year of spent fuel to La Hague for reprocessing. The extracted plutonium (8.5 tonnes) is then shipped to the Melox plant where it is fabricated into about 100 tonnes of MOX. Cadarache has a capacity of 40 tonnes per year and fabricates MOX for France and Germany. Dessel, also at 40 tonnes per year, marketed by AREVA, sells to European and Japanese utilities.

The MOX fuel behaves very well in the reactor, and no failures have been attributed to its use. (Burn-ups of up to 50 GWd/tHM can be obtained using MOX fuel.[203]) The AREVA reprocessing plant at La Hague was commissioned in 1966 with an initial capacity of 400 tons per year. It has since increased capacity to 1,700 tons per year to meet growing customer demand. Additional facilities are being constructed on the same site, including a compaction facility, as well as a plutonium purification and conditioning plant, which was commissioned in 2001. The treatment and conditioning of ultimate residue waste is accomplished by reducing its volume and radio-toxicity to a minimum. The ultimate waste is conditioned in canisters that are internationally licensed for transport and storage. These are stored temporarily at AREVA sites to be transported back to their specified home country for final disposition.

Therefore, the conclusion can be reached that the more reactors loaded with MOX, the less plutonium generated. It takes approximately ten years to complete the fuel cycle back-end process.

In France, MOX assemblies make up 30% of the core of the reactor, the other 70% are assemblies containing enriched uranium. These assemblies are shaped exactly like uranium assemblies, but, in order to optimize the behavior of the fuel, the plutonium content in each rod differs according to its position in the assembly.

The operational results of MOX fuel in a light water reactor are positive. In-reactor behavior of MOX fuel is comparable to enriched uranium fuel. About 30 European reactors already use MOX fuel, a number continuing to increase (EDF plans to have half its reactors operate with MOX fuel within a few years). In Japan, the loading program indicates that by the year 2010 about ten reactors will be loaded with MOX. Studies are being carried out in various countries to design reactors capable of burning 100% MOX fuel. Recycling with MOX fuel reduces waste volume and toxicity from the first recycling; spent fuel elements to be controlled are reduced by 8%, and multiple recyclings reduce plutonium contained in ultimate waste by nearly 100%.

Plutonium recycling with MOX fuel conserves large amounts of natural uranium, and if nuclear power generation capacity is increased, it will conserve coal, natural gas, and oil as well. It is important also that some plutonium byproducts formed during reactor operations, such as $Pu_{240,}$ reduce the effectiveness of plutonium as a potential bomb material by forcing more difficult separation of the Pu_{239} isotope. The irradiation of nuclear fuel generates low, medium, and high level radioactive wastes, with half-lives from a few days to several million years. Reprocessing spent nuclear fuel, which sorts out the different components, simplifies the conditioning and storage options for each type of radioactive waste. The process has proved to be both efficient and safe.

Safety is a primary consideration in the recycling industry from plant design to plant operations. From the safety standpoint, a recycling plant is very different from a power reactor in that it operates at low temperature and low pressure, and it uses static processes lasting for long periods. In the recycling process, there are two predominant safety factors — containment and cooling. Recycling and transforming the nuclear materials recovered into nuclear fuel to generate electricity reduces the magnitude of the waste disposal problem. It reduces significantly the amount of high-level waste ultimately destined to the geologic repository. Clearly, the less high-level waste destined for long-term storage, the lower the risk to succeeding generations of Americans. An additional significant benefit accrues to national security by reducing the likelihood that nuclear materials will find their way into the hands of terrorist groups.

The safety records of the plants are excellent. The spreading of radioactive materials is prevented by a double containment system, wherein a cascade of pressure differentials ensures dynamic containment. The radioactive doses to workers decrease to very low levels because remote handling becomes a normal operation routine. Some United States and Russian excess weapons-grade plutonium is planned to be used in MOX fuel. In the United States the oxalate process has been selected for the proposed "Plutonium Finishing Plant," which is based on AREVA proven technologies. The continuous precipitation and filtration of plutonium oxalate benefit from decades of experience. It has a reliable high throughput of high quality product with small remotely operated equipment. Any waste process, where a precipitation step is projected, can benefit from this technology. It is easy to extrapolate, as criticality exerts stringent size constraints on the oxalate process that would not be required by any substance other than plutonium or enriched uranium.[204] In summary, the handling of highly radioactive solid materials asks for fully remote operation technologies. The continuous dissolvers and their satellite equipment do dissolve nuclear fuels, but they also separate and classify highly radioactive solids. Routing to and from storage with tight containment is a fully remote operation. System maintenance is fully automated, as well.

The Energy Department has declared 55 tons of weapons-usable plutonium to be surplus to national security needs. The legacy of the arms race has presented the Department with an enormous clean-up task. It must dispose of these 55 tons of plutonium. It can do so by pursuing an approach that will render the plutonium inaccessible and unattractive for future weapons applications. With its appalling dependence on foreign oil, it is imperative that the United States design, construct, and operate mixed oxide (MOX) fuel fabrication facilities that will prepare excess weapons' plutonium for irradiation in MOX reactors.

The Department currently stores these plutonium-containing materials at the following sites: Pantex Plant, Rocky Flats (Environmental Technology), Savannah River, Hanford, and the Idaho, Oak Ridge, Los Alamos, and Lawrence Livermore National Laboratories.[205] The more reactors fueled with MOX, the smaller the amount of plutonium stored.[206] From a non-proliferation standpoint, MOX fuel is safer than enriched uranium fuel.

In conclusion, spent fuel should be recycled to recover the unused U_{235} and plutonium produced. Recycling with MOX fuel controls plutonium inventories, because it consumes plutonium and prevents new production. Electricity is generated at a constant or even declining plutonium balance — the more reactors "moxified," the lower the plutonium production.

Vitrification

Wastes isolated during reprocessing are conditioned on-line at the plant as they are separated or produced. Fission products (e.g., strontium and cesium) and transuranics (e.g., neptunium and americium) are formed while the fuel is in the reactor. Once isolated, they concentrate practically all the radioactivity of the spent fuel in a small volume. These non-recyclable products make up the final wastes. They are incorporated into glass that is then cast into stainless steel, leak-tight containers. The containers are then put into ventilated pits before being returned to their owners, which are the utilities of various countries. Vitrification permanently immobilizes the fission products, because glass is insoluble in water and remains inert in contact with natural physico-chemical agents. The structural materials of the fuel, also non-recyclable, are classified as high-level wastes. Standardization of the materials and containers facilitate handling, transport, and storage operations.

In France, the Commissariat à l'Energie Atomique (CEA), the French Atomic Energy Commission, began research on the immobilization of high-level waste (HLW) in 1957. The matrices considered were crystalline materials, phosphate glasses, and borosilicate glasses. Because of their amorphous structure, *glasses* appeared to be capable of incorporating most of the fission product oxides in their vitreous network. By the mid 1960s, borosilicate glasses were selected for the vitrification of HLW solutions as the best compromise in terms of containment (resistance to leach, irradiation, and thermal stability), technological feasibility, and cost, via the volume reduction factor. Today, borosilicate glass has become the worldwide standard and has been chosen for almost all vitrification processes of HLW solutions.

Research and development (R&D) performed by CEA led to the choice of a two-step vitrification process. The vitrified waste is obtained by first evaporating and calcining the nitric acid feed solution containing the fission products. The calcine is then fed together with glass "frit" (materials used in making glass) into an induction-heated metal melter.[207] The basic principles leading to the choice of the two-step process were:

- Separation of the functions; and
- Remote maintenance of the process equipment.

The separation of the process functions (calcinations and vitrification) led to simpler and more compact equipment, which is always desirable in a highly radioactive environment. Easy remote maintenance of all process equipment allowed for complete in-cell assembly and disassembly with moderate size overhead cranes, master-slave manipulators, and remote controlled tools.

Using the two-step vitrification process, France has been operating industrial HLW vitrification facilities successfully since 1978 in the Marcoule Vitrification Facility (AVM). The process was implemented on a larger scale in the late 1980s in the R7 and T7 facilities at the La Hague reprocessing plant (one stop shopping). Both of these plants have three processing lines with a glass production capacity of 25 kg/h. A third facility, using this same process, was commissioned at another plant that began in 1990.

The program has had two objectives: containment of long-lived fission products, and the reduction of the final volume of waste. This is a very much-needed capability in the nuclear process. The feedback from hot operations and long-term R&D programs conducted by the CEA has helped to improve the process in all of its glass formulation processes, associated technologies, and operation and maintenance (O&M). The R7 and T7 vitrification facilities operating in-line with France's two major 800-ton capacity commercial reprocessing plants have had outstanding records of operation, especially in the total glass production and plant availability areas, and with respect to safety, remote in-cell maintainability, and secondary waste generation.

France's long-term commitment to nuclear is evident in its R&D programs that enable her to have continuous improvement in her nuclear capabilities. The average melter lifetime now exceeds the basis value by a factor of two (4,000 hours instead of 2,000 hours). Another development, implemented in 1996 after ten years of operation at La Hague, is the use of mechanical stirring in the melter to deal with higher noble metals content and increase the capacity of the vitrification lines.

By 2002, the R7 and T7 facilities at La Hague had produced more than 11,000 high-level glass canisters, representing more than 4,500 tons of glass and 4.5 billion curies. In-line vitrification of HLW produced by operating recycling plants has become a real commercial industry because in 1995 the first return was made to its customers. The CEA and AREVA COGEMA have expanded on the Cold Crucible Melter (CCM) technology to improve the performance of their vitrification lines at the R7 and T7 plants. The high specific power transferred to the melt will allow high operating temperatures without any impact on the process equipment. The AREVA COGEMA plans to implement the technology in La Hague for the vitrification of highly concentrated and corrosive solutions produced by uranium and molybdenum fuel reprocessing. The Advanced Cold Crucible Melter (ACCM), developed by Cogema, allows an increase in the melting capacity, and the United States could utilize that process for their various high-level waste (HLW) streams.

In the future a major milestone in the evolution of the process will be the deployment of the Cold Crucible Melter (CCM) technology in the R7 plant. The technology will be applied to the vitrification of highly concentrated and corrosive solutions produced by uranium and molybdenum fuel reprocessing. The deployment of the CCM at La Hague will illustrate France's CCM advantages for the immobilization of hard to process waste:

- High operating temperatures;
- Flexibility with respect to waste composition;
- High waste loading factors while maintaining outstanding product quality; and
- Compact design and virtually unlimited equipment service life.

The final (HLW and MLW [mid-level]) reprocessing residues, for example, the non-reusable part of the fuel (vitrified fission products and concrete-encapsulated or compacted hulls and end-fittings), are packaged under internationally agreed specifications. Approval of specifications came on August 1, 1996. The waste and residue, consisting of fission products, hulls, and end-fittings, were approved in France, Japan, Germany, Belgium, Switzerland, and Netherlands. The United States is vitrifying HLW at West Valley and the Savannah River Site, utilizing Horne borosilicate glass. Vitrification of HLW is at INL and Hanford, as well.

According to Dr. Lake, "Three issues should be considered in choosing the best nuclear fuel cycle for the United States in the 21st Century. They are economics, the environment, and social responsibility. Because his questions go straight to the heart of this paper, they are included for consideration. In economics, he poses the following questions:

- Q. *How does the cost of recycling spent nuclear fuel compare with the cost of using it only once before disposal?*

  A. "According to the French data, considering all factors from mining to disposal, the cost of recycling spent fuel is roughly the same as for using it once and then storing it permanently." In other words, it should be a trade off. However, "reprocessing has a potential bonus: There is up to100 times more energy potential in nuclear fuel than is extracted in one cycle. Multiple recycling in future advanced reactors could offer significant advantages in sustaining low-cost nuclear fuel supplies for many generations," *if* nuclear power is expanded.

- Q. *Is it better to directly dispose of all spent nuclear fuel in a geological repository that has been designed to provide barriers to the release of hazardous materials for thousands of years or to reprocess it, recycling the plutonium and other fuel materials in the reactor and disposing only the shorter-lived fission products?*

  A. "Certainly recycling can simplify waste disposal. There would be far less plutonium and other long-lived materials to store in a repository. Multiple recycling, to extract the greatest amount of energy from the fuel and minimize the net waste volume would require fast burner reactors, which was cancelled in the 1970s."

- Q. *Is not separated plutonium more vulnerable to theft?*

  A. "No nuclear materials have ever been proliferated from commercial spent   fuel because of the substantial cost and technical difficulty and the strict oversight by the International Atomic Energy Agency (IAEA)."

Dr. Lake further stated that nuclear power, in one or more advanced states, holds great promise for the generation of abundant clean and affordable electricity for the United States and the world. If spent fuel recycling proves to be cost-effective, and if advanced technology can respond effectively to the proliferation concern, then the United States has an attractive new option, promising clean energy, improved social acceptability, and reduced cost of waste disposal.

Summary

The Idaho National Laboratory (INL) has developed a cheaper and cleaner nuclear waste reprocessing system than those currently in use in the United States or abroad. Utilizing this new technology to reprocess and reuse nuclear materials, the final waste product has a half-life of 30 years and a storage requirement of only 300 years. Most important, the amount of nuclear waste to be stored is reduced by 95%, and that waste can be stored on-site.

The nuclear waste problem can be solved.

CHAPTER 9

U.S. Nuclear Investment and Operating Costs

What has passed may be a lesson.
Thomas Jefferson[208]

For more than three decades, beginning with the Manhattan Project in World War II in the 1940s and continuing into the 1970s, the United States led the world in nuclear research and application both for the defense of our nation and for the good of mankind. Then, paradoxically because he was a nuclear-trained officer in the United States Navy, President Jimmy Carter initiated the downward spiral of U.S. global nuclear leadership by terminating important nuclear projects — *reprocessing* for example — and abandoning research in emerging technologies. To further exacerbate the problem, in the fall of 1977, Carter, together with his ineffective administration and a feeble Congress, created yet another bureaucracy, the Department of Energy (DOE) that fractured the proven infrastructure of the U.S. nuclear community by dividing responsibilities and authorities. Then, because of an accident at Three Mile Island (TMI) in March of 1979, which posed no threat to the public, the public withdrew its support for "nuclear," and no new initiatives were undertaken for the ensuing 30-odd years.

However, one does not hesitate to use nuclear medical instruments in the time of physical pain or faced with a life or death situation, *but* we refuse to accept nuclear power to generate electricity. (In fact, the United States relies on other nations to provide the bulk of our medical isotopes.) The author wanted the reader to see and appreciate the amount of money that the United States has invested in nuclear weaponry to keep us safe as a nation. Even though this is essentially a "sunk cost," we have an opportunity to utilize the nuclear byproducts of the Cold War to generate electricity — to turn waste into something useful, fuel for reactors.

Although the principles of engineering economic analysis would dictate that the U.S. investment in nuclear weapons systems should be considered a sunk cost and thus ignored, it is suggested that the magnitude of the investment is such that the significant residual value resident in spent fuel and deactivated nuclear weaponry should be recovered and put to productive use — and this is possible through recycling and utilization of these resources as fuel for electricity-producing nuclear reactors. Major components of the investment that should be considered are as follows:

- The cost of the Manhattan Project through August 1945, was $20 billion (Note: all costs quoted are *then-year dollars*).
- The total number of nuclear missiles built from 1951-present is in excess of 67,500.
- The total number and types of nuclear warhead and bombs built from 1945-1990, were more than 70,000 in 65 types.

- The number of nuclear warheads requested by the Army in 1956-1957, was 151,000.
- The number in the stockpile as of 1997 was 12,500 (8750 active, 2,500 hedge and contingency stockpile, with 1,250 awaiting disassembly).
- The projected U.S. nuclear warheads and bombs after completion of the START II reduction in 2003 would be 5,000.
- The amount of plutonium still in weapons is 43 metric tons.
- The number of dismantled plutonium "pits" stored at the Pantex Plant, Amarillo, Texas, is 20,000.[209]
- The number of high-level radioactive waste tanks in Washington, Idaho, and South Carolina is 239.
- The volume in cubic meters of radioactive waste resulting from weapons activities is 104 million.
- The estimated 1998 spending on all U.S. nuclear weapons and weapons-related programs was $35.1 billion.

Additional expenditures that must be included are as follows:
- Additional warheads the military wants to hold in active reserve to "hedge" against future threats are 2,500.
- The largest and smallest nuclear bombs ever deployed: B17/ B24 (~42,000 lbs., 10-15 megatons); W54 (51 lbs., 0.01 kilotons, 0.02 kilotons-1 kiloton).
- The peak number of operating domestic uranium mines in 1955 was 925.
- The fissile material produced was 104 metric tons of plutonium and 994 metric tons of highly enriched uranium.

- The number of thermometers that could be filled with mercury used to produce lithium-6 at the Oak Ridge National Laboratory is 11 billion.

- The states with the largest number of nuclear weapons are New Mexico with 2,450, Georgia with 2,000, Washington with 1,685, Nevada with 1,350, and North Dakota with 1,140.

- The total known land area occupied by U.S. nuclear weapons bases and facilities is 15,654 square miles. (The total combined land area of the District of Columbia, Massachusetts, and New Jersey is 15,357 square miles.)

- The legal fees paid by the Department of Energy to fight lawsuits from workers and private citizens concerning nuclear weapons production and testing activities from October 1990-March 1995, are $97 million.

- The money paid by the State Department to Japan following "fallout" from the 1954 "Bravo" test is $15.3 million.

- The monetary and non-monetary compensation paid by the United States to Marshallese Islanders since 1956 to redress damages from nuclear testing is at least $759 million.

- The money paid to U.S. citizens under the Radiation Exposure and Compensation Act of 1990, as of January 13, 1998, is approximately $225 million. There have been 6,336 claims approved and 3,156 claims denied.

- The total cost of the Aircraft Nuclear Propulsion (ANP) program from 1946-1961, was $7 billion. The total number of nuclear-powered aircraft and airplane hangers built has been 0 and 1.

- The number of secret Presidential Emergency Facilities built for use during and after a nuclear war is in excess of 75.
- The currency stored until 1988 by the Federal Reserve at its Mount Pony facility for use after a nuclear war is in excess of $2B. The amount of silver in tons once used at the Oak Ridge National Laboratory, Tennessee, Y-12 Plant for electrical magnet coils is 14,700.
- The total number of U.S. nuclear weapons tested from 1945-1992, is 1,030. (1,125 nuclear devices detonated).
- The first nuclear test was the "Trinity," July 16, 1945. The last one was the "Divider," September 23, 1992.
- The estimated amount spent between October 1, 1992-October 1, 1995, on nuclear testing activities was $1.2 billion with 0 tests.
- The cost of 1946 Operation Crossroads weapons tests "Able" and "Baker" at Bikini Atoll was $1.3 billion.
- The largest U.S. explosion, 15 Megatons on March 1, 1954, was "Bravo."
- The number of islands in Eniwetok atoll vaporized by the November 1, 1952 "Mike" H-bomb test was 1.
- The number of nuclear tests in the Pacific is 106.
- The number of U.S. nuclear tests in Nevada is 911.
- The number of nuclear weapons tested in Alaska, Colorado, Mississippi, and New Mexico is 10.
- The number of naval nuclear propulsion reactors operational is 104 and the number of operational commercial power reactors is the same, 104.
- The current number of attack (SSN) submarines is 80 and ballistic missile (SSBN) submarines are 18.

- The ballistic missile defense spending in 1965 was $2.2 billion, and in 1995 was $2.6 billion.
- The cost of January 17, 1966, nuclear weapons accident over Palomares, Spain, including two lost aircraft, an extended search and recovery effort, waste disposal in the United States, and settlement claims was $182 million.
- The number of designated targets for U.S. weapons in the Single Integrated Operation Plan (SIOP) in 1976 was 25,000, 1986 was 16,000, 1995 was 2,500.
- The number of U.S. nuclear bombs lost in accidents and never recovered is 11.
- The number of Department of Energy federal employees in 2004 was 16,100 and the number of its contractor employees was 100,000.
- The minimum number of classified pages estimated to be in the Department of Energy's possession is 280 million.
- The average cost per warhead to the United States to help Kazakhstan dismantle 104 SS-18 ICBMs carrying more than 1,000 warheads is $70,000.

The total investment probably is incalculable, but the above data give one a sense of its magnitude. Residual values in the nuclear materials should be put to productive use rather than thrown away as nuclear waste assigned to long-term storage, the costs of which in the far future also are incalculable. Thus "value added" to the United States by recycling and reusing these material is such that it no longer can be ignored.

Nuclear weapons belong to the Department of Defense (DoD); nuclear materials are DOE's responsibility. Laboratories have resources to help DOE with planning our energy policies that support its missions. Defense Laboratories report to DoD and DOE, but the President makes the decisions. At one time, Vice President Gore and

Senator Domenici wanted DOE's Defense Programs moved to DoD, but that did not come to pass. The DOE's National Nuclear Security Administration (NNSA) has three deputies: Defense Programs, Naval Reactors, and Nuclear Energy.

The DOE Nuclear Weapons Complex produces and maintains the U.S. nuclear arsenal. The facilities are as follows:

- Lawrence Livermore National Laboratory, California, specializes in stockpile stewardship and maintenance, arms control, and waste management. It employs approximately 6,600 people with an annual budget of $1.6 billion.

- Los Alamos National Laboratory, New Mexico, specializes in stockpile stewardship and maintenance, arms control, and waste management. It employs approximately 9,000 people with an annual budget of $1.8 billion.

- Kansas City Plant, Kansas, specializes in production of weapons. It employs 2,700 people with an annual budget of $400 million.

- Y-12 Plant Oak Ridge National Laboratory, Tennessee, specializes in production of nuclear materials, arms control, and disposal. It employs approximately 6,000 people with an annual budget of $865 million.

- Savannah River Site, South Carolina, specializes in production of nuclear materials. It employs 11,000 people with an annual budget of approximately $2 billion.

- Pantex Plant, Texas, specializes in assembly, dismantling, and disposal of weapons. It employs approximately 3,300 employees with an annual budget of about $540 million.

- Sandia National Laboratories, New Mexico, specializes in weapons design, arms control, and waste management. It employs approximately 8,700 people with an annual budget of $1.4 billion.

- Nevada Test Site, Nevada, specializes in stockpile stewardship and maintenance, nuclear testing, and disposal. It employs approximately 3,500 people with an annual budget of $419 million.

The national investment is huge. Capital investment required to construct new nuclear plants is large as well. But it is important to look at the longer-term life cycle costs associated with nuclear power and compare those costs with the other major power sources, namely coal and natural gas. With oil approaching $150 a barrel in July 2008, it probably should be included in subsequent analyses.

In 2005, the Nuclear Energy Agency (NEA) and the International Energy Agency (IEA) published their sixth report in a series of studies on projected costs of electricity generation. The study was conducted by a group of experts from 19 member countries and two international organizations, the International Atomic Energy Agency (IAEA) and the European Commission (EC). This report presents and analyzes projected costs of generating electricity using data provided by participating experts and generic assumptions adopted by the study group. The levelized lifetime cost methodology was applied to estimate generation costs for more than 100 power plants, including coal-fired, gas-fired, and nuclear. All plants studied rely on technologies available today and considered candidates for commissioning by 2010-2015 or earlier.

Assumptions for the principal technical and economic parameters included an economic lifetime of 40 years, an average load factor for base-load plants of 85%, and discount rates of 5% and 10%. Electricity generation costs calculated are generation costs at the station. They do not include transmission and distribution costs. Costs associated with residual emissions are not included because they are not borne by electricity producers. In view of the risks in competitive markets, investors tend to favor less capital-intensive technologies. However, the methodology adopted for calculating generation costs in this study did not specifically take this factor into account. The markets for natural gas are undergoing substantial changes, and environmental policy influences fossil fuel prices significantly. Security of energy supply remains a concern and may be reflected in government policies affecting future investment.

Given the above considerations, the study finds that the lowest levelized costs of generating electricity from traditional generation technologies and energy sources may be expected to fall within the range of 25-45 U.S. dollars per megawatt hour (USD/MWh) in most countries. The levelized costs of technologies are sensitive to the discount rate and the projected prices of fuel, coal, natural gas, and nuclear. The study provides insights on the relative costs of generating technologies. Within its framework and limitations, the study suggests that none of the traditional electricity generating technologies can be expected to be the cheapest in all situations. The preferred generating technology will depend on the specific circumstances of each project.

Note from the data shown in the Table 10 that although the initial investment is higher than coal or gas, nuclear power is competitive over the 40-year life cycle used in the study, and the capacity factor is above 85%. Given that reactors in the United States can be extended to 60 years, nuclear becomes even more competitive, because the discount rate affects only the capital investment. Table 10 compares costs of the top three producers of electricity.

Members of the Nuclear Power Joint Fact Finding (NJFF) reviewed a number of studies that evaluated the life cycle levelized future cost of nuclear power. Using their own spread sheet analysis and considering the sensitivity of cost and price to certain variable factors, they found that a reasonable range for the expected levelized cost of nuclear power is from 8 to 11 cents per kilowatt-hour (kWh). These estimates are not cost to the consumer, but cost to produce and deliver to the grid. See Table 11.

Table 10 **Range of Levelized Costs***
for Coal, Gas, and Nuclear Plants

	Investment	O&M	Fuel	Total Cost
At 5% Discount Rate				
Coal	8-14	4-10	9-28	22-49
Gas	4-12	1-5	30-45	39-57
Nuclear	10-18	6-10	4-8	23-36
At 10% Discount Rate				
Coal	14-26	4-10	9-28	28-58
Gas	7-17	1-5	30-45	43-58
Nuclear	19-38	6-10	4-8	31-53

*Note: Costs given in U.S. Dollars per Megawatt Hours (MWh).

Table 11 **Summary of Levelized Costs***

Cost Category	Low Case	High Case
Capital Cost	4.6	6.2
Fuel	1.3	1.7
Fixed O&M	1.9	2.7
Variable O&M	0.5	0.5
Total (Levelized Cents/kWh)	8.3	11.1

*Levelized life-cycle cost is the total cost of a project from construction to retirement and decommissioning, expressed in present dollars then spread evenly over the useful life of the reactor.

The cost of recycling or reprocessing spent nuclear fuel, compared with the cost of using it only once before going to disposal, is difficult to determine. First, one cannot calculate the ultimate cost (or determine the acceptability) of the Yucca Mountain geologic repository, which must last thousands of years. Secondly, the United States has no solid data on in-country commercial reprocessing costs. Also, one has to consider the role of the United States Enrichment Corporation (USEC) and the amount of money it cost the United States to subsidize USEC's venture in Russia. The $8 billion deal was made to convert 500MT of highly enriched uranium (HEU) from dismantled Russian nuclear warheads into low enriched uranium (LEU) suitable for use in U.S. commercial reactors. As previously mentioned, USEC is developing an advanced technology plant to increase its capability to reprocess spent fuel to reduce our dependence on imports that account for about 77% of the U.S. nuclear fuel market. It is an energy security question. *It is the same principle as not being dependent on Middle Eastern countries for our oil.*

The United States essentially places zero value on recycled plutonium, which reflects a very short-sighted view, considering the fact that *one gram of plutonium 239 (Pu_{239}) generates as much electricity as one ton* of oil* (*the equivalent of 250-333 gallons, depending on density of the oil). In MOX fuel, the fissile material, plutonium, takes the place of uranium-235 used in fresh fuel.

Tangible considerations must be taken into account, as well. Tangibles are all assets, revenues, expenses, and liabilities that are *measurable*. That is to say, a dollar value can be placed on *tangible* assets. The cost analysis model should be made on a life-cycle basis, and the recommended life is 60 years; therefore, the analysis period should be 60 years as well. This criterion is based on the experience of the French, British, and Americans where reactors originally designed and justified on the basis of a 20 year expected life have been extended to 60 years. The system acquisition costs, or *first costs,* of these plants have long since been amortized. The economical life is very important in analyses that compare nuclear with other options, because the initial construction, equipment acquisition, and setup costs for nuclear plants are significantly higher than similar costs for coal, gas, and oil fueled plants. The operating costs, however, are just the opposite — nuclear plant operation and maintenance (O&M) costs are far less than those for the alternatives mentioned. These observations were borne out in the Nuclear Energy Agency (NEA) and the International Energy Agency (IEA) study referenced earlier.

Reasonable cost estimates for the system cannot be made at this time for the following reasons:

- There are too many major policy decisions yet to be made that result in too many unknowns.
- There are too many additional unknowns and variables that affect the outcome.
- The scale and scope of the problem require a team of analysts to conduct the cost study.
- A realistic cost analysis is beyond the scope of this research.

Although a meaningful benefit-cost analysis of this system is quite impossible at this point, the evidence supporting the concept is so powerful that positive and aggressive actions should be taken now. Finance costs are a major factor to the decision maker, but financing the system should be considered separately and independently from any system cost analysis used to determine which if any alternative should be selected. That decision should be based strictly on the relative merits of the competing systems alternatives.

It is important to recognize that reactor size, hence output, is very important. There is a threshold below which a reactor will not be economically feasible. That threshold is not defined at this point — advanced reactors under development range in output from 50 to 1,000 megawatts. A case can be made that the optimum size reactor for the system envisioned is on the order of 300 megawatts plus or minus 10 to 15%. Much smaller, the reactor is probably not cost-effective — much larger, the reactor is simply too large to meet the requirements of flexibility of siting, modularity, underground installation, standardized design, factory production, and simplicity of operation. The requirement for underground installation is driven by the need to reduce the vulnerability of the plant to sabotage and/ or terrorist attack. Newer modular reactors with low output of approximately 100 – 150 MW can be clustered so that, although they are small, a cluster of four becomes 400 – 600 MW, and operating efficiency is greatly increased. The greatest savings accrue from reduced staffing requirements — basically, by centralizing the control system, one crew can operate the cluster.

Take the nuclear materials recovered from weapons no longer needed, add the recoverable nuclear materials from commercial reactors' spent fuel, process and fabricate new fuel for new nuclear reactors, and produce electric power for the good of us all. Nuclear power generation is, in fact, environmentally friendly, efficient, and safe, as evidenced by successful long-term experience by both the U.S. Navy and France.

CHAPTER 10

Notional Nuclear Power System (NNPS)
Definition and Analysis Methodology

Knowledge must come through
action; you can have no test
which is not fanciful, save by trial.

Sophocles[210]

This chapter describes the notional or theoretical nuclear power system developed as the preferred vehicle by which the hypothesis can be proved or disproved. It defines the power generation model, identifies four feasible alternatives, and describes the planning and implementation of the Notional Nuclear Power System (NNPS) designed to solve the nuclear waste disposal problem. Advanced nuclear reactors are introduced and desired characteristics specified for production, spent fuel reprocessing facilities are described, the alternatives defined, program management structure specified, and program implementation planned by phases.

The methodology used for the analysis consisted of establishing criteria against which the alternatives and candidate reactors were evaluated. Using a Delphi[211] technique, expert value judgments were solicited from professional nuclear engineers and scientists. Multivariate data analysis was the statistical tool employed to conduct sensitivity tests. Collected data were used, not only to provide information but also to conduct sensitivity tests, to determine the consequences of changes in the weights (importance) assigned to the issues.

NNPS Definition

This section addresses the NNPS model, alternatives, planning, and implementation of the nuclear power generation system structured for this research. Note that the electric utilities industry consists basically of three major subsystems: generation, transmission, and distribution. This research effort deals solely with the generation subsystem. Because the transmission lines and distribution grids are in place nationwide and used regardless of whether generation is by fossil fuels or nuclear power, these subsystems are not included in the scope of this investigation.

NNPS Model

The four components of the NNPS are the following:

- Advanced reactors;
- Reprocessing;
- Mixed oxide (MOX) reactor fuel fabrication; and
- Vitrification of radioactive waste.

The "front-end" of the nuclear fuel cycle activities, the mining and milling of uranium ore, are not at issue. Each of the four components is addressed in the following sections with emphasis placed on the two most important elements: reactors and reprocessing. The French and U.S. Navy experiences in developing and operating exceptionally successful nuclear power systems, albeit for quite different purposes, serve as models for the development, implementation, and management of the NNPS. (The U.S. industry is successful also in generation.)

Over the course of the past 20 to 40 years, advances in nuclear technologies have provided solutions to scientists and engineers in many of the problems that plagued earlier generations of reactors. Two of the most significant of these are the *meltdown proof* safety feature and the greatly *increased efficiency* of the advanced reactors under consideration.

In addition to the requirement to reprocess spent fuel is the equally important requirement to burn the plutonium recovered from deactivated nuclear weapons and extracted from spent fuel. The amount of plutonium from the two sources is very different. Surplus "weapons-grade" plutonium (WPu_{239}) is essentially pure, typically about 93% Pu_{239}. When the purity is in the 80-93% Pu_{239} range, the plutonium is referred to as "fuel-grade," and plutonium with less than 80% Pu_{239} is known as "reactor-grade."[212] On the other hand, the amount of plutonium extracted from spent fuel is only about 1% of the spent fuel material, but it needs no enrichment.[213] Plutonium from both sources can be processed, fabricated in MOX plants, and used in reactors to generate electricity. The final objective is to use the reactors to consume plutonium completely, thus avoiding the requirement for geologic disposal altogether.

Advanced Reactor Model

Borrowing from the U.S. Navy's experience and expertise in developing and operating robust, reliable, effective, and safe nuclear reactors, the following characteristics should be included in the specifications for production of advanced reactors for the utility industry in the United States. The desired characteristics are as follows:

- Capable and reliable;

- Standardized to the maximum extent feasible;
- Safe for the environment, the public, and the crews that operate them;
- Positioned on the leading edge of technology, especially in the areas of operational safety and reliability;
- Resistant to shock induced by storm, earthquake, and physical assault, and able to continue to operate or resume operation quickly;
- Resilient to withstand years of frequent power demand changes;
- Designed to be operated and maintained by well-trained but minimal crew;
- Simple in design, small in size, and uncomplicated in operation;
- Inherently safe and can respond to operational transients without the need for immediate operator action;
- Heavily shielded so that plant operators receive much less radiation exposure from the reactor during prolonged exposure than they would receive from background radiation in their daily off-duty lives; and
- Amenable to underground installation — out of sight, and resistant to attack.

Subsequent to the preparation of the specifications listed above, a Department of Energy (DOE) "technology roadmap" for advanced reactor development was found that emphasizes many of the same characteristics.[214]   Reinforcing desired design characteristics for the NNPS enumerated above, they are as follows:

Area I Sustainability

- Sustainable energy generation, clean air, long-term availability of systems, and effective fuel utilization;

- Minimize nuclear waste; reduce long-term stewardship burden, thereby improving protection for the public health and the environment; and
- Increase the assurance that they are a very unattractive and least desirable route for diversion or theft of weapons-usable materials.

Area II Safety and Reliability

- Excel in safety and reliability;
- Have a very low likelihood and degree of reactor core damage; and
- Eliminate the need for offsite emergency response.

Area III Economics

- Have a clear life-cycle cost advantage over other energy sources; and
- Have a level of financial risk comparable to other energy projects.[215]

NNPS Reprocessing

Reprocessing takes spent fuel through chemical processes that recover unused nuclear materials (uranium and plutonium). *Recycling* is the reusing of those materials — the uranium and plutonium again refuel the reactor. The reprocessing complex must have the capability to recover uranium and plutonium from spent fuel. These recovered nuclear materials must meet the specifications required for the receiving fuel fabrication plants that recycle these materials to reactor fuel. Uranium and plutonium recovered from deactivated nuclear weapons can also be recycled to make reactor fuel. At least two reprocessing facilities may be desirable to simplify logistics and to reduce shipping

— one on the East coast and one in the West. This would be an economic decision.

NNPS Mixed Oxide (MOX) Fuel and Fabrication

Savannah River has two MOX fuel fabrication plants that have the capability of fashioning fuel into rods or pellets required to supply the AP600 and CANDU reactors. The facilities could also be modified to produce plutonium or high-enriched uranium (HEU) fuel for the Gas Turbine-Modular Helium Reactor (GT-MHR) and Pebble Bed Modular Reactor (PBMR). A plant having similar capabilities will be required in the Western support facility and, because of its proximity to the Pantex Plant, it must have the capability of fabricating fuel for the Pu-burning advanced reactor selected for the NNPS.

NNPS Vitrification

The concept calls for recycling reactor fuel, particularly fuels fabricated from weapons Pu, to reduce the amount of radioactive materials and toxicity that eventually must be vitrified and placed in a geological repository. Vitrification capability remains a requirement for disposal of fission products. Plants are needed at both East and West support facilities. Current vitrification plants utilize a process whereby nuclear wastes are mixed into molten borosilicate glass, which is then poured into large stainless steel canisters and sealed. There appears to be no reason to specify any criteria for the new facilities beyond their present capabilities.

NNPS Alternatives

Beginning in the fall of 2000, dialogue was established with a network of nuclear experts, who were professional colleagues in the field. Before the literature review was undertaken, informal brainstorming sessions regarding the nuclear waste issue and possible alternative resolutions led to the conclusion that four alternatives were feasible. They fell within the perceived scope of the investigation and could contribute to the resolution of the nuclear issues to varying degrees. The state-of-the-art nuclear technologies required must be available and operating by the year 2020. Major NNPS components, the reactors, reprocessing, fuel fabrication, and vitrification plants, must be affordable, efficient, reliable, easily operated, and safe.

Several other alternatives considered were ruled out, because they could not make a significant contribution toward resolution of the issues being addressed. Among those alternatives suggested but discarded were the following:

- Canada Deuterium Uranium Reactor (CANDU), Advanced Gas-cooled Reactor (AGR), and ABB Combustion Engineering's System 80+ (PWR)
  - o CANDU and AGR are foreign reactors and could pose problems in this proof-of-concept test program
  - o "Evolutionary" designs that fail to meet NNPS specifications.[216]

Coal, oil, natural gas, and hydro as energy sources
  - o Sources are judged, for differing reasons, to be inadequate to contribute significantly to resolution of issues being addressed
    - ▪ Coal: environmental issues;
    - ▪ Oil: supply and cost issues; and
    - ▪ Hydro: high cost and environmental and geological issues.

Renewable energy, primarily wind and solar
  - o Cannot supply significant amounts of energy on a reliable basis
    - ▪ Combined, renewable energy sources provide

less than 1% of power required

- Unpredictable and/or unreliable, theses renewable energy sources cannot be depended on 24/7.

The four alternatives selected, Do-nothing,[217] Modernize existing plants, Utilize selected nuclear power plants as test beds for NNPS, and Construct a state-of-the-art system of advanced reactors, are defined in the following sections.

Alternatives

1. *Do-Nothing*

By definition the do-nothing alternative means continue the present *modus operandi*. No new nuclear plants would be constructed. Operators of the 104 nuclear power plants currently licensed in the United States would continue their efforts to improve efficiency, reliability, safety, and to extend licensed plant life. Basically, these plants would continue to conduct "business as usual" through their remaining licensed life.

2. *Modernize Existing Plants*

This alternative encompasses major modifications and upgrades to increase output, efficiency, safety, and extend the service life of the 104 operating nuclear power plants. The Advanced Boiling Water Reactor (ABWR) and AP600, which are operational Generation III technology reactors, could be used to augment existing plants' output.

3. *Utilize Selected Nuclear Power Sites as Test Beds for the Notional Nuclear Power System (NNPS).*

This alternative provides an operational test program, as a proof-of-concept vehicle, to permit evaluation of advanced reactors and the viability of full system implementation. Initially, the reprocessing element of the system could be tested, and this could be implemented in the near term. Selected reactors would be modified, as necessary, to burn MOX fuel produced at the Savannah River Site (SRS). Two potential locations for this operational test are Oconee, South Carolina, and Palo Verde, Arizona. Palo Verde has three light water reactors (LWRs) that are capable of using 100% MOX cores,[218, 219] and SRS currently is working with Duke Power on a MOX fuel project with the objective of using MOX fuel at its Oconee plant. As soon as the advanced reactors are operationally ready, one or more would be installed at or proximate to the designated proof-of-concept test facility.

4. Construct a State-of-the-art System of Advanced Reactors

This alternative requires construction of the power generation system, utilizing advanced reactors to burn MOX from reprocessed spent fuel, weapons-grade plutonium, and plutonium (which may be "moxified") to increase national electrical power generation capacities. The transmission and distribution subsystems are upgraded under other initiatives; therefore, they are not addressed in this discussion of the alternatives.

NNPS Planning

The scale and scope of this nuclear initiative would involve several government agencies and facilities and require the commitment of serious funding; therefore, it will have to be a government-led and government-sponsored program.[220] The framework of the problem can be viewed in the same light as other major U.S. government acquisition programs. The more significant aspects are addressed in the following paragraphs.

NNPS Steering Committee

To succeed the program must have a *champion* at the highest level of government. It could well be the President; if not, the next most senior of the officials having interest in and responsibility for commercial energy programs. At the direction of the champion or sponsor, a Steering Committee would be formed to provide guidance and oversight of the program. The committee members appointed would be senior executives, one each from the agency responsible for commercial energy programs, the lead laboratory (probably SRS), one or more of the other national laboratories, as appropriate, and one or more of the major nuclear power utility companies, such as Duke Power (the Carolinas) and Dominion Virginia Power (Virginia). There would likely be others as well; however, there would not be representation from any potential competing reactor vendor.

NNPS Program Office

Because the SRS has the nuclear expertise and facilities necessary to provide program support, the Program Office would likely be established there and operate from that site. The appointed Program Manager would be responsible for planning, organizing, and staffing the office, and this could very well be patterned after the Naval Nuclear Propulsion Program (NNPP).

The outstanding success of the NNPP rests on strong, central, technical, and responsible leadership with thorough training, conservatism in design and operating practices, and a complete knowledge of every aspect of the program. Rickover taught his crew well, and he maintained that excellence must be the norm in the NNPP. In order to maintain these standards, individuals must accept responsibility for their actions.[221] Applying these precepts, the career professional managers and operators of the NNPS for the commercial nuclear power industry would be given the authority commensurate with their responsibilities — they would be held accountable.

A set of technical performance metrics must be identified and defined by the program office to measure reactor and system efficiency, safety, and performance. Although there are voluminous performance indicators and metrics available across the complex, there is no standard for utilization, format, content, et cetera. The relationship between the plant operators and DOE is often adversarial, because of its zeal for "standardization."[222] (It is not clear why DOE should be involved in an area seemingly under the purview of NRC.) Because this is an issue, metrics will have to be agreed upon by all the stakeholders prior to commencement of Phase II.

NNPS Implementation

Implementation of the full-scale system is expected to take at least 15 years. The milestones will be addressed in subsequent paragraphs. When fully implemented, the system could serve two markets in most need of additional electrical power: first, the Northeast, with New York and Boston being heavy-demand areas subject to power shortages during periods of unusually high usage, and second, the West Coast, particularly California, which had major power outages in 2000 and 2001.

To facilitate operations in two areas so distant from one another, it may be economical to establish two core service support facilities, each equipped to provide reactor fuel to the system's nuclear power plants. The East Coast choice is essentially predetermined; SRS has the facilities in place to reprocess spent fuel and fabricate reactor fuel in configurations required for the specific reactor(s) selected. West Coast support facility selection is not quite as clear-cut; four locations emerged as being the most likely candidates for selection: Palo Verde

Nuclear Generation Plant, Arizona; DOE Hanford Site, Washington; Pantex Plant Site, Amarillo, Texas; and, INL, Idaho.

Experience in implementing major systems dictates the prudence of establishing a schedule that is deliberate, conservative, and achievable. Milestones must be established that measure progress against plan and provide a timetable for making major program decisions. These are "gates" through which program elements must pass in order to proceed to the next.

To facilitate description and understanding of the proposed evolution, program phases are established to coincide with the milestones; Milestone 1 signals completion of Phase 1, and so it proceeds. Major phases and milestones are described in the following paragraphs.

Phase I: Prototype Demonstration—Milestone 1

The AP600, AP1000, and ABWR reactors are operational. The GT-MHR is underway in Russia, but the program is years behind its earlier schedule. Start-up should be accomplished in approximately 15 years. The PBMR should have an earlier start-up. To allow for unforeseen development delays, Milestone 1 is established as 2025, at which time the two advanced reactors being considered should have operating prototypes.

Phase II: Proof-of-Concept Demonstration—Milestone 2

Because the ABWR, BWR, AP600, and AP1000 reactors are currently operational, the proof-of-concept demonstration applies to the GT-MHR and PBMR. This phase provides the opportunity for each reactor to be constructed and to demonstrate its operability, efficiency, reliability, maintainability, and safety. This will provide a measure of the reactor's *utility* to the industry. Does it perform as intended? Is it reliable? Are there any unanticipated problems? Does it promise siting flexibility? Does it utilize the demonstration plant to provide valuable data for analysis and evaluation of the candidate reactors? This demonstration phase will provide valuable data relative to design, performance, safety, cost, and ease of construction for use in evaluating and analyzing the candidate reactors. Milestone 2 is established at five years subsequent to passing Milestone 1, or by 2030.

Phase III: Manufacture of Standardized Reactor— Milestone 3

This phase would run concurrently, at least in part, with Phase II. It has been suggested that five reactors of approved design would have to be manufactured to determine with reasonable certainty the acquisition costs of the candidate. The need for multiple units is that there is a learning curve in any new industrial or manufacturing process that must be experienced to achieve an acceptable level of efficiency. Ideally, this phase could be completed two to three years after completion of Phase II. Conservatively, Milestone 3 should be realized by 2033.

Phase IV: Implementation of Regional System (East)— Milestone 4

The number of advanced reactors to be acquired, the timing of acquisition, the siting, management, and oversight responsibilities will be determined at such time as the government-industry partnership has been effected, the system defined, and funding identified. The only element of the system that does not currently exist is the reactor. Reprocessing and fuel fabrications could be accomplished at existing SRS facilities. The memoranda of understanding (MOU) with utility companies other than Duke Power would have to take effect, but this should not pose an obstacle. Milestone 4 would be dependent on the scope of the Phase IV program, the production rate of the reactors, and the efficiency of the business and licensing processes. The U.S. nuclear infrastructure, manufacturing, and manpower must be strengthened, and the licensing process streamlined. Milestone 4, to be determined (TBD).

Phase V: Construction of Nuclear Park (West)— Milestone 5

This phase is dependent upon success in the East. Also, implementation of this phase could be unnecessary if government reprocessing and fuel fabrication facilities become operational at Hanford or the Idaho National Laboratory. The NNPS, without use of government reprocessing and fuel fabrication, calls for construction of facilities co-located with or proximate to a major nuclear power facility. At this time the Palo Verde Plant in Arizona appears to be the most likely candidate for the nuclear park. Milestone 5, TBD.

Phase VI: Operation of Regional System (West)— Milestone 6

All of the elements addressed in Phase IV apply. Implementation of the plan and introduction of the regional system in the West will replicate that in the East. However, it should be carried out more efficiently given the benefit of lessons learned from experience in the East. The advanced reactors are well suited to meet electrical demands in remote areas. Milestone 6, TBD.

Analysis Methodology

Given the nature of the problem and the data available, it was determined that the most workable process was to conduct a quasi-statistical analysis[223] of the alternatives. And the first step in the process was to identify and select those feasible alternatives and candidate reactors for evaluation and establish criteria against which they could be measured. The methodology selected to test the hypothesis and answer the eight questions in support thereof (see last paragraph, page 25, and questions on page 26) was to structure a notional state-of-the-art nuclear power generation system for comparison with current, on-line nuclear power plants.

Primary consideration was given to economics, tangible and intangible factors, and the consequences thereof. Using interviews, surveys, and the Delphi technique[224] modified to the extent that communications were direct and accomplished by teleconference and e-mail, expert value judgments were solicited from nuclear engineers and scientists in the civilian and government energy and defense communities, including the national laboratories. The objective was to prioritize the importance of the six national issues being addressed and to evaluate the capabilities of the four alternatives to contribute to the resolution of those issues. Based on these expert value judgments and using a quasi-statistical analysis technique, it was possible to determine an order-of-merit solution that identified the best alternative.

The methodology was structured to embrace the parameters of the problem being addressed and provide a means whereby the alternatives could be evaluated relative to each other. It was not derived from any model found in the literature. This analysis, which determines the relative capabilities of alternative nuclear power systems to contribute to the resolution of specific national issues, has not been found in the open literature. For these reasons it is believed to constitute original research.

The foundational literature search was followed by frequent discussions and question and answer (Q&A) sessions with a small number of experts (5 to 10) in the nuclear field. Using the Delphi technique, consensus emerged on the judgments regarding weights assigned to issues. A questionnaire was distributed to a second group to determine the relative capabilities of the four alternatives to contribute toward resolution of those issues. Each participant is a professional who has worked in the field for an average of 25 years.

This process has experienced four distinct phases and, because of the sensitivity of the topic to world events, the relative importance of the national issues has changed:

- Prior to 9/11, geologic disposal of nuclear waste, high cost of electricity, and dependence upon foreign oil were the top three of seven issues. Immediately following 9/11, homeland security became the most important issue, followed by dependence upon foreign oil, acquisition of nuclear materials by terrorists, and geologic disposal of nuclear waste.
- Six months later, safety emerged as the most important issue, followed by geologic disposal of nuclear waste, and homeland security.
- One year later, safety and geologic disposal of nuclear waste remained numbers one and two, and acquisition of nuclear materials by terrorists replaced homeland security as number three.
- U.S. dependence on foreign sources and high price of oil, exacerbated by adverse effects thereof on the economy, has made dependence on foreign oil *the critical national issue in 2007 and 2008.*

Multivariate data analysis (statistical techniques used to analyze observed values that arise from more than one variable) provided the means to derive information from the data collected. The procedure applies as follows: one dimension being rated (the alternative) and its mate (in this case, the issue), according to its relative importance (weight), are cross-multiplied, then summed for each alternative. This procedure provides a weighted score, and the totals can be compared directly — the higher the rating, the higher the capability of the alternative to contribute to the resolution of the issues. This technique has proved successful across a range of statistical applications, including industrial and commercial analyses.[225]

Research for this book has not been complimented with funding for applied research, testing, and/or fabrication. Research and analysis were tempered by experience and working knowledge gleaned from participation in technical, planning, and strategy conferences, and symposia attended over a period of 15 years. Sponsored by the Departments of Defense and Energy, academia, and industry, the topical issues discussed included: national energy issues, nuclear science and technology, security and infiltration defenses of nuclear sites, industrial nuclear efficiency, proliferation of nuclear materials, transportation of nuclear materials and waste, energy efficiency, nuclear reactor efficiency, nuclear cleanup and redevelopment, and geologic disposal of nuclear waste.

Summary

This chapter defined the NNPS, developed the feasible alternatives, and described the program management structure, and implementation phasing. The methodology utilized in the analysis was presented in detail sufficient to permit replication by future researchers. The next chapter will present the analysis and evaluation of the NNPS alternatives and the candidate nuclear reactors.

CHAPTER 11

Analysis and Results

Wisdom outweighs any wealth.
Sophocles[226]

Overview

This chapter addresses economics, tangible measures, and intangible considerations pertinent to the analysis. The *tangibles* include costs associated with investment, operation and maintenance (O&M), fuel cycle, and decommissioning and decontamination (D&D). The *intangibles* include consideration of non-physical factors, such as dependency on foreign oil, acquisition of nuclear materials by terrorists, pollution of the environment by fossil fuel burning power plants, and other adverse effects on society. Today (2008), with prices as high as $147 per barrel, the cost of oil has to be considered as a tangible variable in the economic analysis, and the adverse effects on society of the oil issue (cost and dependency on foreign sources) have to be considered in the evaluation of the intangibles.[227]

The four alternatives and the Generation III and IV reactors are described, and the advantages and disadvantages identified. Finally, a quasi-statistical analysis is presented that establishes a measure of worth for each alternative. The issues considered are the following:

- Dependence on foreign oil;
- Safety in design and operations;
- Geologic disposal of nuclear waste;
- Acquisition by terrorists of nuclear materials;
- Environmental pollution; and
- Shortage and high cost of electricity.

Analysis and Evaluation of Alternatives

In addition to the discussions of economics, tangible measures, and intangible considerations, this section presents analyses of the four alternatives, the Generation III and IV reactors, and a determination of their relative order of merit.

Economics

Economic competitiveness is a requirement of the marketplace and is essential for Generation IV nuclear energy systems. The transition from regulated to deregulated energy markets is taking place worldwide, creating new opportunities for independent power producers and merchant owner-operators. This means that there will be more options for nuclear power. If the Notional Nuclear Power System (NNPS) being considered proves feasible, Alternative 4 could be one such option. Utilizing state-of-the-art technologies, it should be safe, reliable, and more efficient. Being smaller and inherently safe, Generation IV plants would offer simplicity of operation and flexibility of deployment not possible with the current systems.

To be competitive in the open market, the life-cycle costs of the Generation IV system must demonstrate a monetary advantage. To get to that point government support will almost certainly be required. There are four elements that are generally considered in analyzing life-cycle costs of nuclear power systems: acquisition or capital, operation and maintenance (O&M), fuel cycle, and decommissioning and decontamination (D&D). As with any system or project other factors enter the equation, such as the expected economic life or duration of the project, the rate of acquisition of the major components and associated construction costs, and the financing arrangements for the project. The rate of acquisition determines the rate of production of the reactors. This variable has a huge impact on equipment costs — economies of scale can make or break a project. In the following sections major considerations in the analysis and evaluation of the project are addressed.

Tangible Considerations

Tangible considerations take into account all assets, revenues, expenses, and liabilities that are *measurable,* that is to say a dollar value can be placed thereon. The cost analysis model should be made on a life-cycle basis. Although some nuclear reactors in the United States will operate for 60 years, the recommended analysis period is 30 years, which is often used in studies that address the economics of nuclear power.[228] Based on the experience of the French, British, Japanese, and U.S. nuclear power industries that brought nuclear power stations on line in the 1950s and 60s and are operating today, this criterion is considered conservative. Many reactors in France, England, and the United States have been operating for more than 40 years; several have been cleared for 60.[229]

The majority of these reactors were originally designed and justified on the basis of a 20 year expected life. The system acquisition costs or *first costs* of many of them have long since been amortized. This statement raises the question, "What about *stranded costs?*" Consider them in the following context. An economically significant consequence to the nuclear power industry of deregulation is stranded costs, which are defined as "the decline in the value of electricity-generating assets due to restructuring of the industry." Two excellent sources of information on the subject are Gail Cohen and Bruce Biewald. Note that their articles were prepared before the fact, so they are predictive in content, but much of what was foreseen is coming to pass. Under deregulation, prices of electricity have risen; therefore, the nuclear industry is not experiencing stranded costs. People need electricity, and they are willing to pay for it.

The Congressional Budget Office (October 1998) report offers the following conclusions: ". . . the transition to a more competitive electricity industry will be especially costly for many current suppliers."[230] Also, "On this issue, economic efficiency plays second fiddle to fairness and politics." Finally, ". . . the decision to compensate — to ease the burden of restructuring on the owners and creditors of utilities — is ultimately one for the regulators and legislators."[231]

The economic life is very important in analyses comparing various nuclear with other energy options, because the initial construction, equipment acquisition, and setup costs for nuclear plants are significantly higher than similar costs for coal, gas, and oil fueled plants. The operating costs are just the opposite — nuclear plant O&M costs are generally less than those for the alternative energy sources mentioned. The average non-fuel O&M cost for a nuclear power plant in 2007 was 1.29 cents/kWh.[232]

The Electric Power Research Institute (EPRI), a nuclear lobbying group in Washington, DC, has estimated costs; electric utilities that are going to build or buy plants also have estimated costs. However, reasonable cost estimates for the NNPS cannot be made at this time for the following reasons:

- There are too many major energy policy decisions yet to be made.
- There are too many unknowns and too many variables that affect the outcome. (To date, no figures have been tabulated for a closed fuel cycle with reprocessing.)[233]

Cost estimating is the critical element. The NNPS is a complex activity that involves several high-technology facilities and the skills of many highly trained and experienced scientists and engineers. Time is also an important factor in the equation; the shorter the deadline, the larger the team required to prepare the estimate. These factors drive the need for detailed planning, scheduling, estimating, and managing the activity. Components of the estimate include direct and indirect costs, equipment acquisition and materials, skills and labor rates, subcontracts, travel and transportation, services, labor and material burdens, and administrative expenses.

Simply collecting the necessary data poses a huge problem, which is compounded by uncertainties of future costs of reactors that are still in the prototype development phase. Stewart describes cost estimating as a process involving the following steps:[234]

1. Define the process;
2. Develop the methodology and estimate schedule;
3. Formulate a project schedule;
4. Prepare cost estimate ground rules;
5. Estimate labor hours and materials and other direct costs;
6. Review, compile, organize, reconcile, and compute the estimate;
7. Analyze the resulting estimate for credibility, competitiveness, completeness, comprehension, and

compatibility with budget constraints;

8. Revise, if required, and publish; and

9. Use the estimate.

Benefit-cost analyses are conducted to determine, from the economic or *best value* point of view, the best of the alternatives being considered. Often, however, other considerations may be overriding in making the final decision. Financing, for example, although separate from the engineering economy analysis, will drive the final decision, because interest on construction loans to build the plants is the greatest cost.

In the case of the Gas Turbine-Modular Helium Reactor (GT-MHR), the optimum output, which drives the physical size, was determined by the developer to be 285 megawatts electric (MWe). As the design output decreases toward 100 MW, the cost-effectiveness decreases, and it becomes less competitive. When the design size increases above 285 MWe, modular production becomes infeasible. Major factors influencing the competitiveness of the GT-MHR are its standardized design and the fact that it can be mass-produced in a factory.[235]

Compared to the cost of production by coal and gas, major competitors, the cost-effectiveness of new nuclear power plants is more sensitive to discount rates and amortization periods. (Finance repayment structures are a separate issue.) The principal contributing factors to the sensitivity are the larger acquisition costs associated with the nuclear power plant and the discount rate used in the analysis; however, the acquisition cost differential is decreasing as more stringent pollution controls for fossil plants are being mandated.

"The most recent Organization for Economic Cooperation and Development (OECD) comparative study shows that at a 5% discount rate, in 7 of 13 countries considering nuclear energy, it would be the preferred choice for new base-load capacity commissioned by 2010. At a 10% discount rate the advantage over coal would be maintained in only France, Russia, and China."[236]

Benefit-cost analyses of nuclear power show it to be competitive even against coal, which is the fuel of choice in 52% of power generation plants in the United States. The U.S. Utility Data Institute figures for 2007 show nuclear at 1.68 cents per kilowatt-hour (kWh), down from 2005 cost of 1.72 cents.[237] Comparative 2005 costs of oil at 8.09 cents, gas at 7.51 cents, and coal at 2.21 cents make the nuclear alternative very attractive. Some 1,034 nuclear plants are included in the database.[238] As costs of gas and oil have increased over the past decade, costs of nuclear power have steadily decreased. Year 2000 figures, for example, show higher costs for gas, and lower costs for nuclear power because of a robust 4% increase in output from existing nuclear plant capacity.[239]

It is especially important to note that most studies have made no attempt to quantify the eventual costs to society resulting from environmental pollution and depletion of natural resources, consequences of using fossil fuels. Air pollution, a major problem in today's society, does considerable damage to our environment and to human health.[240] The major types of air pollution are the following:

- *Gaseous pollutants*: A different mix of vapors and gaseous air pollutants is found in outdoor and indoor environments. Most are carbon dioxide, carbon monoxide, hydrocarbons, nitrogen oxides, sulfur oxides and ozone. The most commonly recognized type of air pollution is smog.

- The *Greenhouse effect* prevents the sun's heat from rising out of the atmosphere and flowing back into space. This warms the earth's surface, causing the green house effect. Activities, such as the burning of fossil fuels, are creating a gaseous layer too dense to allow the heat to escape.

- *Acid rain* forms when moisture in the air interacts with nitrogen oxide and sulfur dioxide released by factories, power plants, and motor vehicles that burn coal or oil. This interaction of gases with water vapor forms sulfuric acid and nitric acids. Eventually, these chemicals fall to earth as dew, drizzle, fog, snow or rain.

Damages accruing from the effects of acid rain are currently being evaluated in the courts because of lawsuits filed by certain eastern states against utility companies in the Midwest that operate coal-fired plants.[241]

When construction, acquisition, and operating costs, including fuel of Generation IV reactors, can be more clearly defined, conducting a meaningful cost-effectiveness analysis, using any one of several cost analysis models, will be feasible. One such analysis vehicle is DECARTES, a project that developed databases and software to be used in comparative assessment studies.[242] Also, the Los Alamos National Laboratory (LANL) has developed a comprehensive collection of economic analysis programs and techniques that are useful in evaluating energy and environmental issues, including monetary and financial flows analysis, natural resource and energy economics, environmental analysis, and engineering economics.[243] Sandia National Laboratories (SNL) developed a methodology called "Consequence Based Analysis for Infrastructure Surety" that addresses risk and reliability issues.[244] Small cost differentials between nuclear and coal, its closest competitor, lead to the conclusion that implementation of this system will be determined by one or several over-riding intangible considerations (factors that cannot be monetarily quantified). Although priority of these intangibles may change from time to time, at the moment of this snapshot, summer of 2008, a good case can be made that United States dependency on foreign oil is a very high priority. The situation had been developing over the past year with the price of oil reaching its apparent peak in July. (Note that each one of the alternatives discussed reduces our dependency on foreign oil. Were it not for the criticality and broad effects across society of this dependency, the order of priority would be: reactor safety, geologic disposal of nuclear waste, and acquisition of nuclear materials by terrorists.)

The capabilities to reprocess spent fuel and burn plutonium increase the capacity to generate electricity and do so in environmentally friendly advanced reactors. The bottom line is that costs in terms of potential lives saved, destruction or damage to ecology and atmosphere are simply undeterminable. There is no way to predict the dollar value of the consequences of not reprocessing and burning the nuclear waste.

Investment Costs

As previously stated, investment costs depend on political and energy policy decisions yet to be made and final definition of the system to be implemented. The IEA *World Energy Outlook* 2006 compares projected (2015) nuclear, coal, gas and wind electricity costs assuming nuclear investment costs of $2,000/kW and $2,500/kW. Other key figures include projected investment costs for coal of $1,400/kW and gas of $650/kW. (Costs of nuclear, coal and natural gas of 0.5, 2.2 and 6.0 $/gigajoule [GJ], respectively). Using a low discount rate and investment cost, nuclear is the cheapest option at $49/MWh. At a high discount rate and investment costs, investment in nuclear is $57/MWh, which is cheaper than gas-fired power, but more expensive than coal so long as coal prices do not exceed $70/ton. It should be noted that increases in costs of fuel affect the cost of electricity produced by nuclear the least. A 50% increase in uranium, gas and coal prices would raise the electricity cost by 3% (nuclear), 20% (coal) and 38%

(gas), thus making nuclear the cheapest option, even with a higher capital costs of $2,500/kW.[245]

The prototype Generation IV reactors must be up and running in order to establish the parameters of the analysis and enable analysts to estimate the costs with an acceptable degree of certainty. The design target acquisition cost for Generation IV reactors is $1,000 per kWe. The manufacturer of the GT-MHR has predicted to come in under target, approximately $975 per kWe,[246] and the Pebble Bed Modular Reactor (PBMR) in the 10-module commercial-scale version at the design cost of $1,000 per kWe. The cost of a single module PBMR plant is estimated to be about $1,325 per kWe.[247]

The general rule applicable to the manufacturing process is that the learning curve has a very significant effect. In the initial stage of production of a standardized product, as the number of units produced is doubled, the labor cost of production is reduced by 20%. Labor cost for the 4th reactor produced in a plant would likely cost 80% as much as the labor cost for the 2nd reactor, which represents an "80% learning curve."[248] This rule reflects the experience of the aircraft industry that could serve as a model for mass production of Generation IV reactors.

As to plant or facility construction costs, it can be said that they will vary according to the output of the reactor(s). Likewise, but probably less correlated, will be the physical size of the reactor and its production costs. Prototype Generation IV reactors now being developed will provide a baseline for cost estimates in the near future. Some of the system's cost categories that must be considered are the following:

- Site acquisition and preparation;
- Reactor production and procurement costs that include:
  o Engineering development
  o Manufacture of reactor containment, pressure vessel, and other major components

   o Subsystem component procurements from suppliers;
- Assembly, plumbing, and wiring; and
- Transportation.

Both projects have incentive to meet the Generation IV plant target specification of $1,000 per kilowatt-hours. It would be imprudent to speculate as to the final production cost.

Operation and Maintenance (O&M) Costs

The World Nuclear Organization (WNO) reports that *production costs during the period 1995 – 2005, including O&M and fuel, for nuclear were the lowest of any reliable electricity source* averaging 1.72 cents per kWh, lower than coal at 2.21 cents per kWh gas at 7.51 cents per kWh, and oil at 8.09 cents per kWh . These figures were based on Nuclear Energy Institute (NEI) Global Energy Division data, updated in June 2006. (Note: the above data refer to fuel plus O&M costs only. They exclude capital, since this varies greatly among utilities and states, as well as with the age of the plant.)[249]

Fuel Cycle Costs

One of the most attractive aspects of nuclear power has always been low fuel cost compared to coal, oil, and gas. A fundamental reason is that uranium ore is plentiful and relatively cheap, and there is no competing demand for it. Even with processing, enrichment, and fabrication into reactor fuel, the price in January 2007 was approximately $1,800 to get one kilogram of uranium oxide (UO_2) reactor fuel, which compared to the cost of fuel for a coal-fired plant is typically about one-third and for a gas-fired plant about one-quarter.[250] Projections for the period 2005-2010 show the following comparisons for countries around the globe[251] (see Table 12):

Table 12 Fuel Costs:* Comparison of Coal and Gas with Nuclear (2010)[252]

	Coal	Gas	Nuclear
North America: Canada	3.11	4.00	2.60
United States	2.71	4.67	3.01
Europe: France	3.33	3.92	2.54
Germany	3.52	4.90	2.86
Russia**	3.33	2.93	2.12
Far East: China**	2.00	-	2.83
Japan	4.95	5.21	4.80
Korea	2.16	4.65	2.34

*Cost in cents/kWh; Discount Rate 5%.

**Costs for Russia and China calculated using data from referenced report.[253]

The costs for the "Nuclear" fuel given above are for a "once-through" fuel cycle. Significant cost savings may accrue from reprocessing spent fuel and recovering the plutonium and uranium that are then used in the MOX fuel. Although the reprocessing and fabrication costs are high, they are offset by the facts that MOX needs no enrichment, and the amount of high-level waste produced at the end is greatly reduced. Seven UO_2 fuel assemblies are used to produce one MOX fuel assembly with some high-level waste that must be vitrified.[254] The bottom line (see Chapter 8) is that reprocessing and reusing spent fuel can result in about 95% reduction of the volume, and the requirement for and cost of permanent disposal reduced as well.[255]

As a matter of interest, Canada uses what is called the Levelized Unit Electricity Cost (LUEC) methodology for measuring the costs of existing plants, which focuses on the life-cycle average cost per unit of electricity, taking into account all capital and operating costs. (The LUEC is a means for calculating the cost of a unit of energy, using a methodology that allows one form of technology to be compared with another.)[256] This all-in unit cost can then be compared as one important measure of the relative attractiveness of each investment option. At a 5% discount rate, analysis showed the following (in LUECs): Coal at 43.1; Gas at 44.4; and Nuclear at 34.2 for the CANDU 9 (2x881 MWe) and 39.8 for CANDU 6 (2x665 MWe).[257]

Decommissioning and Decontamination (D&D) Costs

The process of phasing out an industrial facility is part of its life cycle. Although dismantling its non-nuclear parts creates no particular difficulty, dismantling elements in contact with radioactivity requires special techniques for separating and storing components. The D&D costs associated therewith are generally calculated to be in the range of 9 to 15% of the initial capital costs of nuclear plants. In a recent life cycle cost analysis, Shay and colleagues determined their estimate of decommissioning costs of radioactive units to be 10% of the unreduced unit construction costs in dollars of the same year (constant dollars). (They used previous studies of light water reactor [LWR] power plant decommissioning events, as reference data.[258]) A second study estimates decommissioning costs to be in the range of 9 to 15% of the initial capital cost of the plant.[259] When discounted, these costs add a few percent to the investment cost and even less to the generation cost. In the United States, they add only 0.1-0.2 cent per kWh, which is less than 5% of the cost of the electricity produced. The back-end of the fuel cycle adds another 5% to the overall costs per kWh. This figure is less if there is direct disposal of spent fuel.

Intangible Considerations

Because intangibles are non-physical assets, costs usually cannot be assigned. In specific cases, such as highway construction, lives saved by the "safer" highway often are assigned a monetary value and considered benefits in the benefit-cost analysis. In this scenario four intangible considerations appear to be overriding. They are discussed in the following sections.

Terrorists' Acquisition of Nuclear Materials

Acquisition of nuclear materials by terrorists is one of the major concerns of our country today. Rogue-led nations or terrorist groups conceivably could wreak havoc were they able to obtain sufficient plutonium or high-enriched uranium (HEU) for making nuclear weapons. The more likely threat may be that of the "dirty" bomb. Insecure nuclear materials *anywhere* constitute a threat to *everyone, everywhere.* Given the threats, it is reasonable to argue that the costs associated with the acquisition of a system, which accomplishes the following, are costs that must be paid:

- Reduces the amount of waste destined for geologic disposal;
- Reduces accessibility of nuclear materials to rogue-led nations;
- Serves to keep nuclear materials out of the hands of terrorists or groups thereof; and
- Puts nuclear waste to work productively in the generation of needed electricity.

Environmental Impacts

Air pollution, a major issue in most industrialized nations, including the United States, is caused by the introduction into the atmosphere of chemicals, particulate matter, or biological materials emitted by fossil-fuel-burning vehicles, power plants, industrial furnaces, bacteria, viruses, and such. For that reason, the emission of carbon dioxide by fossil fired plants is an important factor to be considered in the fossil fuel power plant versus nuclear power plant comparison. About a kilogram of air pollution is avoided for every kilowatt-hour of electricity generated by nuclear power.[260] Richard Rhodes states, "Among sources for electric power generation, coal is the worst environmental offender. Recent studies at the Harvard School of Public Health indicate that particulates from coal burning are responsible for about 15,000 premature deaths annually in the U. S. alone."[261] Worldwide emissions of carbon dioxide from burning fossil fuel are estimated to be on the order of 25 billion tons per year.

An important adverse consequence of air pollution is acid rain, which threatens human health and damages crops, animals, property, and forests. Acid rain pollutants may also be present in fog and snow even in the dry form of acidic gases and particulates. Virtually all fossil fuel burning power plants in the United States contribute to the problem.[262] A major and concentrated source of this particular industrial pollution is in the Midwest states — coal-burning power and manufacturing companies there are being sued for damages by certain states in the Northeast.[263] Should large settlements be awarded in these lawsuits, the case for nuclear will be even stronger.

On the other hand, emissions from nuclear power plants are minimal. Murray Stewart (President and CEO of the Canadian Nuclear Association) has pointed out that nuclear power generation is zero-emitting with respect to greenhouse gas emissions; no solid wastes are released to the environment, and all minor emissions from nuclear facilities are decidedly *lower than background radiation levels.* (Background radiation level is that level at which it has been proven there are no harmful effects on humans or the environment. Some examples are radiation from the sun, the houses we live in, and other sources frequently encountered, such as X-rays.) On a global basis, nuclear energy reduced carbon dioxide emissions by 1.8 billion tons in 1995. Electric utilities would have emitted 32% more carbon dioxide that year without their nuclear plants.[264]

Dependence on Foreign Oil

According to T. Boone Pickens, "...we must reduce our dependency on foreign oil because it is killing our economy. In 1970, we imported 24% of our oil; in 1990, it was 42%; and in 2008, it was a high 70%. This means that $700 billion is leaving our country each and every year. This figure is four times the cost of the Iraqi War. In addition, the cost of heating oil is going to adversely impact our country this winter."[265]

Society

Environmentalists, public health advocates, and citizens' groups contend that coal-fired plant emissions pose a clear and present danger and that many of them should be shut down or retrofitted to reduce emissions. Is it worth the cost? There is mounting evidence that coal-fired power plants have a devastating impact on public health. Landmark studies by Harvard researchers in 1993 and the American Cancer Society in 1995 documented premature deaths among people chronically exposed to fine particulates of these emissions. Power plants built before 1980 generate half the nation's electricity, but nearly all of the utility industry's unhealthy sulfur dioxide, nitrogen oxide, and soot.[266]

Nationwide, as many as 30,100 deaths a year are related to fossil power plant emissions; a conclusion reached by Abt Associates, a company that does work for the Environmental Protection Agency (EPA). By comparison, ~16,000 Americans are killed each year in accidents involving drunk drivers, and ~17,000 are victims of homicides.[267] A recent study by the Harvard Center for Risk Analysis concluded that fine particle emissions from nine major coal-fired power plants in Illinois were responsible for 400 deaths a year throughout the Midwest, mostly related to heart and lung diseases that could be largely avoided. "Coal-fired power plant pollution is a known trigger for respiratory problems, and not enough is being done to correct the situation."[268] These aging coal-fired power plants are an issue sparking an intense debate over health risks posed by pollution from older plants, how much should be spent to improve or replace them, and whether the benefits derived are worth the cost. Industry officials say there are limits to how much they can spend on antipollution measures and remain competitive.[269]

Tighter controls on power plants could also result each year in 2,000 fewer emergency room visits, 10,000 fewer asthma attacks, and 400,000 fewer incidents of upper respiratory symptoms.[270] In the Chicago area alone, an estimated 400,000 to 500,000 people suffer from asthma. The prevalence of asthma among Chicago area residents exceeds the national average by 25%. With children the asthma rate is more than double the 6 to 7% national average. Pollution from the Chicago power plants and the plants in Waukegan and Peoria affect nine states in the Midwest. Plant operators have spent $200 million to reduce nitrogen oxides (NOx), but they will have to spend hundreds of millions more to install new pollution scrubbers.

A frequently overlooked aspect of energy production is the danger to workers engaged in the process. Consider Table 13, which shows quite clearly the relative safety records of the energy industry by fuel source. Recall, coal-fired power plants provide 52% of our electricity, and nuclear plants approximately 20%.

Analysis of Alternatives

The analysis presents a comparison of the four alternatives under consideration. Examination of the tangible and intangible benefits and relative capabilities offered by the four alternatives provides a meaningful sense of the relative merits of each.

1. *Do-Nothing*

The current national nuclear plan is to continue to operate the in-place plants, D&D reactors as their useful lives end, continue the massive clean-up projects at DOE nuclear waste sites, prepare nuclear waste for final storage, and deposit same in the Yucca Mountain geologic repository. This alternative is costly and risky, because it does not have the capability to burn recycled nuclear materials without modification. It cannot contribute to the resolution of the nuclear waste issue.

Table 13 **Comparative Risk*.**

Source	Fatalities
Coal	342
Natural Gas	85
Oil	418
Nuclear	8

*Based on 1 terawatt-year of energy (A terawatt is 1,000,000 megawatts)

2. *Modernize Existing Plants*

The continued operation of these 104 operating plants (69 pressurized water reactors [PWRs] and 35 boiling water reactors [BWR])[271] is required until many advanced units can be built. Even if licensed for extended life times, they will end their licensed period in the next 15 to 20 years. This power must be replaced. Also, most of these existing plants cannot use MOX fuel unless modified, and modification and service life extension programs would be expensive and short-lived. Although these plants reduce our foreign oil imports, they cannot provide additional reductions. Because this alternative, *Modernize Existing Plants*, does not provide the capability to burn plutonium, unless plants are moxified, it does not meet the established criteria.

3. *Utilize Selected Nuclear Power Sites as Test Beds for the Notional Nuclear Power System (NNPS)*

As mentioned earlier, three of the operating MOX-capable LWRs are located at Palo Verde, Arizona. In the East, it appears that Duke Power's Oconee reactor(s), which are located in South Carolina, will be the first on-line to use MOX fuel. Until the Generation IV reactor is available, this alternative permits testing of the reprocessing and MOX fabrication components of the NNPS. Further, the Generation III reactors are approved and available should increased nuclear capacity be required in the near term, before the Generation IV reactors could be produced in adequate numbers. The capability to burn MOX would contribute toward resolution of the nuclear waste issue, because it consumes reprocessed spent fuel.

Should Generation IV reactors be deployed as small-scale add-ons to existing nuclear plant sites, a distinctively attractive concept in the initial phases of implementation of the NNPS, two significant advantages would accrue: first, an NRC site approval has already been obtained for the current plant, and second, substations and transmission lines are in place. This situation would support realistic proof-of-concept testing, using the GT-MHR and PBMRs.

4. *Construct a State-of-the-art System of Generation IV Reactors*

This alternative requires acquisition of new Generation IV reactors that burn MOX fuel and WPu fuel and implementation of the NNPS envisioned. Small size and standardization of design make it possible to produce in numbers and thereby reducing production costs. This constitutes a leap forward from the LWRs that are individually designed and must be constructed on site.

Flexibility of deployment is feasible, because the GT-MHR reactors may be employed as single units or as multiples. For example, the GT-MHR and the PBMR can be installed in groups; three GT-MHRs or ten PBMRs could be set up to provide approximately 1,000 megawatts, the equivalent of the large LWR plants currently operating in the United States. These plants could be located strategically to provide electricity to those areas of the country most in need of additional electrical power.

The GT-MHR and PBMR are designed to survive the failure of components without fuel damage and without releasing radioactivity, because their fuels can safely withstand maximum temperatures under the worst of circumstances. The GT-MHR design limits the power density of the reactor core, along with the actual size of the core, and exploits natural processes to remove heat and avert fuel damage, even in the event of a total loss of coolant. A major attribute of the Generation IV reactors lies in the fact that they can be installed underground, out of sight, and resistant to attack. Furthermore, they do not produce any atmospheric pollution to the environment. *Every Generation IV reactor that comes on line will reduce our dependency on foreign oil.* Because this alternative does have the capability to burn nuclear waste, it would contribute to the resolution of the nuclear waste issue.

Analysis of Generation III Reactors

This section discusses the Generation III reactors, the AP600, CANDU, ABWR, and System 80+. Since they are licensed and operating, they could be used in the following two roles: first, as a test bed for the system's reprocessing plant's production of MOX fuel, and second, to augment the nation's current nuclear capacity.

1. Advanced Pressurized Reactor (AP600)

The AP600 is a smaller version of the AP1140 reactor that is in use throughout the world. This reactor has the following drawbacks:
- Size: Large (650 Mwe)
  - o Must be constructed on-site;
- Safety: Not passively safe; and
- Security: Requires protection management oversight.

2. Advanced Boiling Water Reactor (ABWR)
- Size: Large (1371 Mwe)
  - o Must be constructed on-site;
- Safety: Not passively safe; and
- Security: Requires protection.

3. System 80+
- Size: Large (1300 Mwe)
  - o Must be constructed on-site;
- Safety: Not passively safe; and
- Security: Requires protection.

4. Canadian Deuterium Uranium (CANDU) Reactor.

The DOE is considering this reactor, but it was eliminated in this study, because the disadvantages inherent in acquiring a foreign-produced reactor override any advantages it might offer. Also, of concern are the following:

- Size: Large (600-700 Mwe)
  - Vulnerable to diversion or theft of nuclear materials;
- Safety: Not passively safe;
- Security: Requires protection
  - *Open borders;* and
- Economic Impact (Adverse on U.S.)
  - Loss of jobs
  - Loss of GNP.

None of these Generation III reactors has all characteristics essential to satisfy the requirements of the NNPS.[272]

Analysis of Generation IV Reactors

The final selection decision regarding Generation IV reactors rests with the utilities that purchase them, or with DOE if a government-led program emerges. The following descriptions provide the results of this analysis, which is based on the best information currently available. Each of the reactors represents state-of-the-art, proven technology and is environmentally friendly.

1. Gas Turbine-Modular Helium Reactor (GT-MHR)

The characteristics of the GT-MHR design offer the following:

- Security: Modular design adaptable to underground installation;
- Flexibility: Burns plutonium from reprocessing or weapons and HEU;
- Size: Most efficient size (325 MWe)

     o   Large enough to produce electricity efficiently; but
- Small enough to be factory produced;
- Affordable: Standardized modules, factory produced;
- Safety: Meltdown proof; and
- Waste: Minimal.[273]

2 Pebble Bed Modular Reactor (PBMR)

   The PBMR characteristics are as follows:

- Security: Modular design adaptable to underground installation;
- Flexibility: Burns plutonium from reprocessing or weapons and HEU;
- Size: Small (180 MWe)
  - o   Too small to produce electricity efficiently
  - o   Must be ganged to attain minimal efficiency of scale; and
- Technology: Refueling online.[274]

Evaluation of Alternatives

The following tables show the results of *weighted* analyses of the four alternatives measured against six national issues.[275] Sources of the inputs are scientists and engineers from Yucca Mountain, Idaho National Laboratory, and the combined U.S. Navy (Naval Station Norfolk and Newport News Shipbuilding) and Utilities (Dominion Virginia Power, Surrey, Virginia). The tables show a simple "ordering of alternatives" based on the capability of each alternative to contribute to the resolution of the national issues addressed. (Tables 14, 15 and 16.)

Table 14 **Yucca Mountain* Weighted Analysis of Alternatives**

Issues	Weight	Alternatives			
		Alt. 1	Alt. 2	Alt. 3	Alt. 4
Dependence on Foreign Oil	5	1.0	2.0	2.0	4.0
Safety in Design and Operations	5	1.8	2.9	3.3	4.7
Nuclear Waste: Geologic Disposal	5	1.9	2.4	2.8	4.2
Acquisition by Terrorists of Nuclear Materials	4	1.9	3.0	2.7	3.4
Pollution of the Environment	2	3.1	3.4	2.9	4.2
Shortage and High Cost of Electricity	1	2.3	3.3	3.3	4.6
Total Score		39.6	58.6	60.4	91.1

* Note: Geologic disposal advocates.

223

Table 15 INL* Weighted Analysis of Alternatives

Issues	Weight	Alternatives			
		Alt. 1	Alt. 2	Alt. 3	Alt. 4
Dependence on Foreign Oil	5	1.0	2.0	2.0	5.0
Safety in Design and Operations	5	1.0	2.2	3.2	5.0
Nuclear Waste: Geologic Disposal	5	2.4	2.8	2.4	4.0
Acquisition by Terrorists of Nuclear Materials	4	1.4	2.8	2.6	3.6
Pollution of the Environment	2	1.4	3.6	3.0	3.2
Shortage and High Cost of Electricity	1	1.4	4.0	2.2	4.2
Total Score		31.8	57.4	56.6	95.0

* Note: Advanced reactor advocates.
They saw no need to go through proof-of-concept testing.

Table 16 U.S. Navy and Utilities* Weighted Analysis of Alternatives

Issues	Weight	Alternatives			
		Alt. 1	Alt. 2	Alt. 3	Alt. 4
Dependence on Foreign Oil	5	1.0	2.0	5.0	5.0
Safety in Design and Operations	5	2.0	3.0	3.8	4.6
Nuclear Waste: Geologic Disposal	5	1.6	2.2	3.2	4.4
Acquisition by Terrorists of Nuclear Materials	4	2.2	2.6	3.0	4.6
Pollution of the Environment	2	2.6	3.0	3.6	4.4
Shortage and High Cost of Electricity	1	2.0	3.0	2.8	4.4
Total Score		39.0	55.4	67.0	101.6

* Note: Advanced reactor advocates.

Using the source data above, the consolidated result is shown in Table 17. Alternative IV dominates in all cases, which effectively negates the meaningfulness of the weights.

Table 17 **Consolidated Weighted Analysis of Alternatives**

Issues	Weight	Alternatives			
		Alt. 1	Alt. 2	Alt. 3	Alt. 4
Dependence on Foreign Oil	5	1.0	2.0	2.0	4.7
Safety in Design and Operations	5	1.6	2.7	3.4	4.8
Nuclear Waste: Geologic Disposal	5	2.0	2.5	2.8	4.2
Acquisition by Terrorists of Nuclear Materials	4	1.8	2.8	2.8	3.8
Pollution of the Environment	2	2.5	3.3	3.1	4.0
Shortage and High Cost of Electricity	1	2.0	3.4	2.9	4.4
Total Score		37.2	57.2	61.3	96.1

The weights assigned range in value from 5 to 1, based on the importance of the issue. For example, the value 5 is assigned to Dependence on Foreign Oil, Safety, and Nuclear Waste, the most important issues, and 1 is assigned to Shortage of Electricity, the least important of the six issues. The value assigned to the capability of the alternative to resolve the issue is shown in the appropriate box. The products of the weight assigned to the issue and the judged capability of the alternative to resolve that issue are summed and shown as the "Total Score." The score provides a number that enables the ordering of the alternatives by desirability, nothing more. The process is as follows:

To get the score, let weights be $a_{1...6}$ and the alternatives $b_{1...6}$ then the score is:

Score Alt. 1 = SUM = $a_1{}^* b_{11} + a_2{}^* b_{12} + a_3{}^* b_{13} + a_4{}^* b_{14} + a_5{}^* b_{15} + a_6{}^* b_{16}$

Thus: Score Alt. 1 = SUM = 5* 1.0 + 5*1.6 + 5*2.0 + 4*1.8 + 2*2.5 + 1*2.0 = 37.0.

Note: the weights assigned are relative, not absolute. Thus, the weight of "1" assigned to *shortage and high cost of electricity* does not mean that the issue is not important, it simply means that, in the ordering of these issues, it is judged to be less important than the others. A survey was then conducted wherein a second group of 20 experts in the civilian and government nuclear and defense communities were asked to judge the capability of each alternative to contribute to the resolution of each issue. Of the 20 asked, 19 responded in time to be included in the analyses.

The total scores show that Alternative 4 with 96.1 outscores the other alternatives by a significant margin. Alternative 4, the state-of-the-art system, scores 3.8 - 4.8 in its ability to contribute to the resolution of each of the issues, and it is judged to be superior in each area. It is important to note that the respondents did not see any significant benefit by opting for Alternative 3, which scores 61.3, only 4.1 points higher than Alternative 2.

Summary

Because the Generation IV reactors are capable of burning plutonium recovered from deactivated weapons, all of the technological capabilities required to meet the proposed objectives have been satisfied.

> *Each of the eight questions pertaining to nuclear waste has been answered.*

CHAPTER 12
Review and Discussion of Issues

Destiny waits alike for the free man
as well as for him enslaved by another's might.
Aeschylus[276]

Prior to the two oil crises of the 1970s, the first in '73 followed by the second in '79, the United States was the world leader in nuclear technology, but that role was aborted by President Jimmie Carter's executive order to terminate all reprocessing activities. And, largely as a result of the nuclear accident at Three Mile Island (TMI), the public adopted an intense anti-nuclear persona, which persisted until fairly recently. Today, however, we see a flurry of activity to resurrect reprocessing, even to look seriously at new construction — and, the French have arrived to help us! Areva is in Bethesda, Maryland!

The eight questions that were raised in Chapter 1 have been thoughtfully considered, and each will be answered in order.

Question 1: *Is technology available to support such an alternative?*
Answer: *Yes.* Technology is available to support such an alternative. Reprocessing involves the separation of plutonium and uranium and other fission products from spent fuel. First, the fuel is *fissioned.* From the spent fuel, plutonium is separated and encapsulated into ceramic-coated particles. The mixed oxide (MOX) fuel is fabricated for the MH-GTR advanced reactor wherein fission energy heats the helium gas that drives the turbine-driven electric generators. The recycling process continues. Ultimately, the remaining waste is encapsulated and poses no proliferation risk. It decays after 300 years, not thousands of years, and is no more toxic than uranium ore that is mined throughout the world. We have the technology.

Question 2: *Does the United States have the capability to utilize available technologies to implement the alternative?*
Answer: *Yes.* The United States does have the capability to utilize available technology to implement the alternative. However, the United States has not reprocessed spent nuclear fuel since the Carter administration prohibited the commercial chemical recycling and separation of plutonium from spent nuclear fuel. After completing the non-proliferation policy review, begun under the Ford administration, the Carter administration cancelled the breeder reactor commercialization and plutonium recycling programs. This action terminated the licensing proceedings, but recycling can be readily reenergized.

Question 3: *Does the alternative source of power help solve one or more national issue?*
Answer: *Yes.* We are in a world where the risks posed by certain rogue nations and other nuclear-capable countries that have

unstable governance are very high. And, terrorist groups pose a real threat, as well. So, homeland defense, nuclear waste, and dependence on foreign oil are the most important national issues, and each can be resolved to a significant degree by recycling. The GT-MHR prototype is being constructed in Russia and will be operational in the near future. The proposed system can be implemented in phases beginning with recycling uranium and fuel fabrication (MOX), followed by installation of the reactors, as they come off the assembly line.

Question 4: *Can the alternative be implemented in the near future?*

Answer: *Yes.* The United States has the capability to utilize available technologies to implement the alternative, if the administration and Congress take the necessary action. Vice President Cheney has said that the U.S. industry must build more than one new power plant per week, every week, for the next 20 years, in order to meet the nation's electricity demands during the next two decades.[277] According to the federal Energy Information Administration (EIA) data, energy demand is expected to grow by almost 50% by 2030. The National Energy Policy (NEP) has recommended at least maintaining and possibly expanding the role of nuclear energy, as a major component of our nation's electricity generating capability. Currently, nuclear power plants generate about 20% of the electricity produced in this country. Just to maintain the 20% share, the United States needs to bring new nuclear power plants on line at a rate of three to four per year, starting in 2015.

Question 5: *Is the implementation of the alternative economically feasible?*

Answer: *Yes.* It is economically feasible, as France's experience with recycling has shown. Therefore, there is no reason to believe that it should not be economical in this country. Furthermore, the economic risk of not implementing the system is unacceptably high. Prices of natural gas have doubled over the past five years and, because of flat supply and increasing demand, are expected to rise by 6% per year. Crude oil prices have been rising continuously since 1998, when the price was $10 a barrel. In July 2008, the price topped $147 a barrel. That is an increase of 1,370% in just ten years. Although the price of coal is not as volatile as the prices of gas and oil, coal prices have been increasing, and the costs of new construction of coal power plants have doubled. Adding to the uncertainty in the coal power sector is the lack of action in Congress to define pollutant emission standards for coal-fired power plants. As these market forces continue to pressure the power industry, nuclear power with its lower fuel costs, lower operations and maintenance costs, and zero pollutant emissions, looks better, and better, and better.

Question 6: *Can the alternative be sustained through the long term?*
Answer: *Yes.* As long as the United States operates nuclear submarines and nuclear power plants, it will have the spent fuel to recycle the fuel in the proposed nuclear power system. Reactors typically generate about 20 metric tons of spent fuel annually. At present, spent fuel is being stored in steel-lined, concrete pools, or in dry containers, because power plants are running out of pool space. At the end of 2006, 60 plants had no more pool space. Even dry storage is currently in use. The United States has to embrace

nuclear power! Recycling recovers the usable plutonium and uranium in its spent fuel.

Question 7: *Will the alternative provide sufficiently large amounts of electrical power reliably and safely in an environmentally sound system?*

Answer: *Yes*. If the necessary funding and support are forthcoming, the alternative will provide sufficiently large amounts of electrical power. Because of the reliability of the nuclear systems of the French commercial experience and the U.S. Naval Nuclear Propulsion Program (NNPP), the safety record of nuclear operations is by far the best of the major alternative sources: coal, gas, and oil. There have been no fatalities, resulting from nuclear accidents or incidents in the history of nuclear operations in the United States. Environmentally, nuclear power is the friendliest of all major energy sources. It is not a polluter of the environment.

Question 8: *Is there a viable alternative to long-term storage of nuclear waste?*

Answer: *Yes*. The United States has a viable alternative to long-term storage of nuclear waste, which is a major obstacle to nuclear acceptability. If this issue were resolved, nuclear power would become the best energy source for the generation of electricity, the friendliest of all sources to the environment, and *provide a major boost to the economy* by lowering the trade deficit and creating thousands of jobs. The gas turbine modular high temperature advanced reactor (GT-MHR) can generate electricity economically, reduce the nuclear waste problem in the process, and frustrate attempts by *would-be evildoers* to obtain nuclear materials.

In 1980, the Reagan administration lifted the ban on domestic recycling, but the world situation was different. The economic and political environments were against the resumption, and it never became an option. President Clinton announced his Nonproliferation and Export Control Policy in a speech at the United Nations on September 27, 1993. A key point of the policy was that fissile material from dismantled nuclear weapons and commercial nuclear spent fuel would be retrievable for at least 50 years or longer in a repository.

Residual value of nuclear materials should be put to use, rather than thrown away in the form of nuclear waste placed in long-term storage, the length of which is *unknown*, the ultimate effects of which are *unknown*, and the ultimate costs of which also are *unknown*. Thus, the *value added* to the United States by recycling and the reuse of these materials is such that the issue can no longer be ignored.

With the world's growing energy requirements, especially in third world countries, they must be dealt with or else they will result in consuming huge amounts of irreplaceable fossil fuels, ripping away forests, and fighting for resources to sustain themselves. Environmentally friendly, state-of-the-art nuclear reactors hold great promise, as an energy source throughout the world. They are twice as efficient, create half the waste, and exhaust virtually nothing harmful to the atmosphere.

General Atomics, a U.S. technology and nuclear energy company, working with Russia, is developing a new generation of nuclear reactors, fueled by millions of tiny, ceramic-coated particles of plutonium removed from Soviet-era nuclear weapons. Russian officials like the idea of recycling. According to General Atomics Vice Chairman, Linden Blue, "They (the Russians) say if we're going to throw away all this, let's at least extract its energy value." The program, which is jointly funded by the United States and Russia, is building a prototype reactor in Siberia that should be operational in about 15 years.

As John L. Sackett, Argonne-West National Laboratory, once stated, "One man's waste is another man's treasure." Democratic Majority Whip Harry Reed, Nevada, liked the concept, because it gave him another reason to argue against the government's $6.7 billion plan to construct a federal repository for high-level wastes in his state, Nevada, at Yucca Mountain. Other proponents agree that the process would largely pay for itself by selling the electricity the Russian government-owned reactors would generate.

As of 2003, U.S. nuclear power plants had produced approximately 49,000 tons of nuclear waste, which is projected to exceed 70,000 tons by 2035. *The current plan is to bury it all in Yucca Mountain*, the nation's long-term waste repository in Nevada, where it will remain dangerously radioactive for much longer than 100,000 years. Under a separation process, plutonium and other heavy elements with the longest radioactive lives would be removed. Ninety-six percent of the separated waste or about 47,000 tons is uranium-238 that would then be removed and placed in storage for low-level nuclear wastes. About 1% or 400 tons of plutonium would be broken down in reactors into less radioactive elements. The wastes from recycling, plus similar wastes from the separation process, a total of 1,960 tons or 4% of the original waste, would be placed in long-term storage in Yucca Mountain.

Vice President Cheney has said that the biggest question still facing nuclear power plants is how to dispose of the nuclear waste they create. In an interview on NBC's "Meet the Press," Cheney described the "ongoing waste problem" and "re-licensing" of old nuclear power plants, as two crucial issues that must be addressed. "If we don't find policy solutions to those issues, then what will happen is we'll have to start shutting down nuclear plants, and then your options will be whether you're going to burn more coal or more natural gas to fire the plants that have to be built to make up for the nuclear plants that go off-line."

The United States simply cannot turn its back on nuclear power, because of the hundreds of tons of uranium and plutonium that exist in surplus weaponry. In the arms agreement, which calls for the destruction of tens of thousands of nuclear weapons, the United States must destroy the uranium and plutonium from these weapons. It can do so by providing much needed electricity. The best way to destroy this plutonium is to burn it in new, safe, melt-down-proof reactors. By placing the plutonium in glass logs and burying it only gets it out of sight — it does not *destroy* it. It is available for chemical separation and potential use by terrorists. No one should be satisfied with any plan that does not take advantage of the huge energy supply. Plutonium can be utilized in nuclear reactors and destroyed for all time. If not, it can fall into the wrong hands.

The best non-proliferation plan is to destroy the plutonium.

President George W. Bush decided to take *the path to disarmament* with President Putin, promising to reciprocate "in kind" to rid the world of 8,000 nuclear warheads. Bush's decision to slash the U.S. strategic arsenal to 1,700 - 2,200 warheads over the next ten years and the Russian response thereto reinvigorated a disarmament process that had stalled out in the decade, following the end of the Cold War.[278] Bush did not want to create a new treaty. The two sides discussed an agreement in which the verification provisions of the first Strategic Arms Reduction Treaty, START I, which was signed in 1991, mandated that each side slice that number to 6,000 with on-site inspections and monitoring of each other's progress. As part of an agreement, following the breakup of the Soviet Union, Ukraine destroyed its final missile silo, and Putin announced that the last nuclear warhead brought to Russia from Ukraine had been destroyed in October 2001. Russia wants to cut even deeper to save money. The United States still had 7,013 warheads in July 2001, but policymakers expected to get to the 6,000 ceiling by December of that year.

In 1993, a second treaty, START II, would cut both arsenals to 3,000 - 3,500 warheads. Although ratified, the treaty has never *entered into force.*[279] Clinton and then-Russian President Boris Yeltsin agreed in 1997 to pursue a START III that would go to 2,000 - 2,500 warheads, but the idea never got off the ground. The Pentagon opposed it, because of a dispute over U.S. missile defense plans. The White House included a figure of 2,200 in deference to those concerns, but preferred the lower number. Because of the September 11 terrorist attacks, the relationship between the two countries has changed. According to Putin, "Terrorists hoped to intimidate us, to take advantage of our differences, to divide us, but what they achieved was our consolidation and solidarity — I would say *solidarity* unheard of in modern history." This does not seem to still be true.

Governments around the world are concerned and are tightening the security of their nuclear facilities, as the International Atomic Energy (IAE) is warning that nuclear terrorism "seems far more likely in the wake of September 11." The United States established "no-fly" zones over civilian and military nuclear installations to prevent a suicide aircraft crash that could release hazardous radiation. France has set up antiaircraft missiles to protect the large pond used to store spent nuclear fuel at its Cap La Hague reprocessing plant. There has never been a case of nuclear terrorism. Although about 400 cases of nuclear smuggling have occurred in the past decade, none involved enough fissionable material to build nuclear weapons. However, if terrorists are willing to die, the security equation changes rapidly. There are three scenarios of serious concern:

- Renegade groups could accumulate material to make a nuclear bomb.

- Terrorists could construct a "dirty" bomb, a conventional plastic explosive or TNT salted with nuclear isotopes that could be detonated to spread radioactivity across a large area. (There are more than 10,000 possible sources of radioactive material around the planet for a terrorist to steal.)

- An airplane or truck bomb breaching the walls of a nuclear plant and releasing toxic radiation poses a very serious risk. There are 438 nuclear-powered electric generating stations in the world, mostly in Europe, and about 250 nuclear fuel plants. The reactors are shielded, and they have safeguards, but terrorism wasn't considered when they were designed. If an airplane or truck bomb hit a reactor or spent fuel pond, it could release radioactive particles into the atmosphere.

Siegfried S. Hecker, former director of the Los Alamos National Laboratory (LANL), said that during the Cold War, the Soviet Union built 20,000 nuclear warheads. Today, although the Russian strategic force is declining, many thousands of warheads remain deployed at dozens of locations and more than 60 storage sites. In addition, 1,000 metric tons of weapons-grade highly enriched uranium and between 125 and 200 metric tons of plutonium are spread throughout the country at various facilities. Russia maintains a large network of production facilities for uranium enrichment and nuclear reactors that continue to produce weapons-grade plutonium (WPu), as well as a network of three-dozen nuclear weapons laboratories and dozens of specialized defense institutes. Hecker warned then that the Russian nuclear complex is largely intact, vastly oversized, and overstaffed. The primary joint program for protection, control, and accounting for nuclear materials and warheads at many of these facilities "has all but come to a standstill." Not only did he blame increased Russian security, but also U.S. bureaucratic demands that have "lost sight that these are Russian nuclear materials in the Russian nuclear complex."[280]

There are two programs to reduce nuclear material, and they have had mixed success. One, which is to turn highly enriched WPu into fuel, has been successful. The other, which is to immobilize plutonium or use it in civilian reactors so it cannot be used for weapons, has never gotten started. The Russians prefer to create electricity from this material. Also, the Russians continue to produce plutonium from reactors they use for energy generation, and they see plutonium as part of their broader plan to encourage nuclear power.

The United States Enrichment Corporation (USEC) includes General Atomics, Lockheed Martin, Allied Signal, Westinghouse, Pleiads Group Limited, and a New York Company with ties to Russia. It was privatized in 1998 at the cost of $3 billion, which it paid to the government. With support from the Bush administration, the outlook for the nuclear power industry brightened. Of first importance is long term purchases of nuclear fuel recycled from Russian nuclear weapons, which is a key part of USEC's revenue stream and is scheduled to eliminate the equivalent of 4,500 Russian nuclear warheads, while contributing $1.7 billion in U.S. currency to Russia. It is proposing eliminating 15,000 more potential warheads in the next 12 years.

"Security is as good as its weakest link, and loose nuclear material in any country is a potential threat to the entire world," said Dr. Mohamed ElBaradei, Director General of the International Atomic Energy Agency (IAEA). A crude nuclear device, a "dirty" bomb, could be detonated by a terrorist group who would have no qualms whatsoever about using such a weapon to wreak havoc on a metropolitan area. Most people are unaware that if highly enriched uranium is at hand, it is a relatively simple job to set off a nuclear explosion.

The IAEA reports that there is a shocking lack of control at nuclear facilities, especially in Russia, to prevent the theft of highly enriched uranium and plutonium. The IAEA cites 175 cases of trafficking in nuclear material since 1993, including 18 cases that involve small amounts of highly enriched uranium or plutonium. Up to 60% of nuclear material in Russia remains inadequately secured, according to Matthew Bunn, Harvard University's Kennedy School. Terrorists may seek to kill scores and cause widespread panic by dusting a conventional bomb with radioactive material. "Suppose that the 19 hijackers had formed into teams to drive four vans with large high-explosive bombs into the power reactors and spent fuel ponds for a large nuclear facility," said Bunn.

According to Roger L. Hagengruber, a Sandia National Laboratory scientist who directed an advisory task force for the Pentagon on unconventional nuclear defense, advanced capabilities exist that could guard against the threat posed by a small nuclear bomb. These include unmanned sensors, radiation detectors, and imaging tools emitting gamma rays and neutrons that can penetrate shields to locate highly enriched uranium. The challenge will be to integrate them in a permanently deployable system, which could be operated cheaply by law enforcement personnel to check ships and trucks at busy harbors and border crossing.

The Pentagon and Defense Science Board estimated that 1,500 tons of weapons capable nuclear material exists in Russia. A weapon could be made from less than four pounds of such material and could weigh as little as 20 pounds delivered by artillery. Worldwide, about 5,000 tons of weapons-capable material is available. Short of a nuclear weapon, the Board said that terrorists might be able to produce a "radiological dispersal device" that would use a conventional explosive to spread toxic plutonium or strontium-90 over a wide area.

Charles Krauthammer wrote, "Here we are, for the second time in a decade, risking American lives in a war against an enemy fueled and fed by oil money. Here we are again decrying our dependence on oil from a particularly unstable, unfriendly part of the world. Here we are again in desperate need of both energy conservation and new energy production. And, fifteen of the murderers on September 11 were Saudi. Their leader is Saudi. Most of their money is Saudi. And that same Saudi money funds the madrassas, the fundamentalist religious schools where poor Pakistani, Afghan, and Arab children are inducted into the world of radical Islam and war against the American infidel. Yet, we bow and scrape to the Saudis. We beg and borrow. We tolerate their deflecting onto America the popular hatred that would otherwise be directed at their own corruption. Why, because, we need their oil. We have known since 1973 that we need to reduce our dependence on Persian Gulf oil, but we have never been serious. It was assumed that Sept 11 would make us serious. We need energy independence."[281]

A recent Associated Press poll reports that 50% of Americans support nuclear power, and 56% of the supporters said that they wouldn't mind a nuclear plant within 10 miles of their own homes.[282] Though support for nuclear power is rising, our national energy strategy will need to address the real threat to nuclear power, besides terrorists, because anti-nuclear activists and politicians have commandeered the nuclear process.

Nuclear energy is a must if we are to become energy independent.

CHAPTER 13

A Vision for Tomorrow

Where there is no vision, people perish.
The Holy Bible, Proverbs 29:18

Conclusions

Electricity is a national resource that cannot be stored. Energy is not only an economic commodity, but it is a political commodity, as well. The demand is up 35% since 1973.[283] By 2025, the Census Bureau predicts the U.S. population will grow about 20% to 337 million.[284] Energy supply must expand by at least 20%.

President George W. Bush approved Yucca Mountain as the repository for high-level waste.[285] It was supposed to be the key to his administration's plan to support nuclear power in the United States. *Aggressive nuclear programs could go a long way in solving major national issues: dependency on foreign oil from the Persian Gulf, long-term storage of spent fuel, potential proliferation of nuclear weapons by rogue nations or use of "dirty" bombs by terrorists, shortage of electricity, and environmental pollution.* Creating a robust nuclear industry is still the right thing to do for the country.

France and the U.S. Navy have provided the examples of excellence after which our nuclear initiatives should be patterned. For 40 years, France has brought to the world her expertise and proven solutions to the complete fuel processes. She has demonstrated her utmost concern for reliability, cost-effectiveness, and safety in all nuclear fuel cycles — front end and back end — covering the entire spectrum. The U.S. Navy Nuclear Propulsion Program (NNPP) has designed, developed, and operated small, reliable, efficient, and safe reactors that have proved successful far beyond the expectations of even the most enthusiastic and optimistic of its supporters. Possibly the best example of its reputation is the fact that its nuclear ships are welcomed in ports around the world, without question as to their nuclear integrity.

Nuclear power has to play a more prominent role in meeting tomorrow's energy challenges. Taking a longer view, the world's energy consumption is predicted to increase 50% by 2030.[286] Fossil energy prices are constantly in a state of flux, and the uncertainty that surrounds the security, supply, and geopolitical areas of production raise serious concerns.

Nuclear power will help solve these problems.

As a result of the oil crisis in 1973, European countries had two options. France chose nuclear and has become the world leader in nuclear power and the closed fuel cycle. Today, the French enjoy the lowest electricity prices and the cleanest environment in Europe. And, France is an energy *exporter.* Italy chose oil and natural gas. Her total power production costs have increased as much as 40% in one year alone (2000), because of the volatile and steadily increasing costs of crude oil and natural gas. And, Italy is an energy importer, relying on foreign sources, primarily France, for over 15% of its energy needs. Italy, the only G8 country without a nuclear power plant, is Europe's largest energy *importer.*[287]

In January 2001, after weeks of threatened power outages, California was hit by rolling electricity blackouts, affecting 500,000 people in San Francisco, Sacramento, and San Jose, as well as other sections of Silicon Valley. The crisis had come to a head the previous summer after San Diego Gas and Electric passed along its wholesale costs to customers, driving up bills by 300%. When the utilities had their own prices capped, they suffered about $12 billion in losses over a six-month period. California stepped in to bail out the largest utilities and re-regulate the system.[288]

California eventually spent $11.7 billion in the space of nine months to purchase energy to halt shortages that had been plaguing the area for almost a year.[289] The problems that California had to endure could have been a great deal more severe if not for the limited nuclear power it had online at the time. Senator Murkowski of Alaska said, *"Nuclear power keeps this nation's electrical grid stable and reliable. . . . Without it, the California grid would have collapsed."*[290]

The required technologies are proven and available to support implementation of a state-of-the-art power generation system; whereby, nuclear materials currently destined for long-term storage can be recycled and utilized productively to generate electricity. Three cases that exemplify effective and safe use of nuclear power are offered as proof of the concept's viability:

- First, the United States Navy had the *foresight* to develop nuclear propulsion that propelled the U.S. fleet to a position of naval supremacy for the past half century;

- Second, a successful and cost-effective commercial venture to generate electricity is exemplified by the French, who had the *will* to go nuclear; and

- Third, commercial nuclear reactors, based on successful Navy designs, generate nearly 85% of the world's nuclear electricity.[291]

These three pillars provide the genesis for the national energy initiative developed herein.

The questions posed in Chapter 1 have been answered in the affirmative.

1. Technology is available to support such an alternative.
2. The United States has the capability to utilize available technologies to implement the alternative.
3. The alternative source of power solves one or more major national issues.
4. The alternative can be implemented in the near future.
5. The implementation of the alternative is economically feasible.
6. The alternative can be sustained through the long term.
7. The alternative will provide sufficiently large amounts of electrical power reliably and safely in an environmentally sound system.
8. Finally, *there is a viable alternative to long-term storage of nuclear waste and continued dependency on foreign oil.*

"Environmental" groups have held nuclear power hostage, even in the face of the severe problems caused by our dependence on foreign oil. The Nuclear Regulatory Commission (NRC) has to be more proactive in its oversight of the industry; in particular, its licensing process has to be more efficient. Our country has the expertise in nuclear power. Why not develop it? Our current reactors have been given an extension on their operating license. The U.S. Navy has trained young men on its nuclear submarines. Let them weigh in on this issue, not the politicians who gave us Yucca Mountain. The utilities should have proceeded, but the Federal government failed to support them. Elected officials wanted to make sure they were re-elected, because their constituents are vague where nuclear is concerned. Global warming wouldn't be a hot issue today had we not relied so heavily on oil and coal. The United States is not solely responsible for global warming. Environmentalists need to talk to China and India.

Recommendations

1. The solution to the problem, facing the United States, is to embrace nuclear energy and enjoy the environmental benefits of recycling its nuclear waste. Is going nuclear more an issue of sovereignty, than economic sense? Thus, it is recommended that the administration vigorously pursue the rejuvenation and modernization of the nation's commercial nuclear power generation industry. The amount of non-reusable waste can be reduced to 1/5 and the toxicity to 1/10. It also allows for the recycling of valuable energy materials, which accounts for 96% of spent fuel. As AREVA's CEO Anne Lauvergeon said, *"Nuclear power is not the only answer, but without it, there is no answer."*[292]

2. Analysis of the advanced reactors expected to be available by 2010 supports recommendation that the Gas Turbine-Modular High Temperature (GT-MHR) be selected as the most promising alternative. Because it is designed to burn pure weapons-grade plutonium or mixed oxide (MOX), it meets the objective of ridding excess weaponry in the United States. At the same time, it eliminates emissions of pollutants to the atmosphere.

It is 50% more efficient than current nuclear plants, because it operates at a much higher temperature.

3. It is also recommended that the Palo Verde plant be designated as the nuclear park for the West coast MOX and fabrication facilities initiative, and the GT-MHR be phased in at the earliest possible date. This would demonstrate proof of concept to the viability of the system.

4. Because the Energy Department has proven inefficient and ineffective as overseer of nuclear energy programs, it is recommended that it be eliminated. A new agency would be formed to manage commercial energy programs, and the Defense Department would be responsible for all aspects of nuclear weaponry and nuclear propulsion, including oversight of the national weapons laboratories.

5. It is recommended, further, that DOE's primary research functions be transferred to universities or the National Science Foundation.

The United States has the knowledge, technical capability, wherewithal, and now it has the opportunity (provided it can muster the *will*) to proceed with a dynamic, robust nuclear power program. The nation cannot afford to pass on this important issue, if we are to become energy independent. *Nuclear power is not the only answer, but without it, there is no answer.*[293] It is recommended that the U.S. energy policy pursue, as a matter of highest priority, the objective of reducing our dependency on foreign oil, and at the same time reducing the quantity of nuclear waste destined for long-term storage. The goal of the DOE Radioactive Waste Management Program should be to minimize the quantity of nuclear waste destined for Yucca Mountain. To attain that goal:

Read, Recycle, Reprocess, and
make a better tomorrow
for Little Joe!

END NOTES

1   Letters of William James (American philosopher and psychologist) to W. Lutoslawski, *Bartlett's Familiar Quotations*, 17th edition (New York: Little Brown and Company, 2002), page 581. item 3. (Hereafter referred to as Bartlett's, 581.3.)

2   "Said What?" http://www.saidwhat.co.uk/quotes/favourite/carl_gustav_jung/we_cannot_ change_anything_unless_we_5546; Internet accessed 5 August 2008.

3   Al Gore (Albert Arnold Gore, Jr., former Vice President), *From red tape to results: creating a government that works better and costs less*. (Washington, DC, Report of the National Performance Review, 7 September 1993), at http://www.ipo.noaa.gov/About/ npr.html; Internet accessed 5 August 2008.

4   Angela Antonelli, *The One Percent Budget Showdown: Clinton's Veto Threats In Perspective* (Washington, DC, The Heritage Foundation, 7 October 1998), at http://www.heritage.org/Research/budget/BG1224.cfm; Internet accessed 5 August 2008.

5   Ronald D. Utt, Ph.D., *Improving Security at DOE Weapons Labs, Executive Memorandum #622* (Washington, DC, The Heritage Foundation, 7 September 1999).

6   Martha M. Hamilton, *Nuclear Power Since Three Mile Island* (Washington, DC: Live Discussion, 1999). http://www.washingtonpost.com/wpsrv/national/talk/archive/ hamilton/0330.htm; Internet accessed 10 February 2003.

7   "U.S. Nuclear Reactors," *Union of concerned Scientists: Nuclear Power Information Tracker,* Cambridge, MA (n.d.) at http://www.ucsusa.org/clean_energy/nuclear_safety/ reactor-map/reactors/three-mile-island-unit-1.html; Internet accessed 5 August 2008.

8   Tim Radford, "Chernobyl death toll under 50," *The Guardian*; Tuesday, 6 September 2005.   http://www.guardian.co.uk/ environment/2005/sep/06/energy.ukraine; Internet accessed 19 July 2008.

9   Edward F. Sproat III, "The Management and Disposal of Spent Nuclear Fuel & High-Level Radioactive Waste in the United States," Presented to: International Conference on Radioactive Waste Disposal, Bern, Switzerland; 16 October 2007 (Washington, DC: Department of Energy), at http://www.icgr2007.org/Proceedings/Session%204/ Presentations/Session4_4_Sproat_Rev.pdf; Internet accessed 19 July 2008.

10   "Spent Nuclear Fuel and High-Level Radioactive Waste Transportation" (Washington, DC, Department of Energy: National Transportation Program (n.d.) http://www.orau. org/PTP/New_ Folder/new/library/DOE/TRANSPORTATION/spentfuel.pdf; Internet accessed 5 August 2008.

11   "Management and Disposition of Excess Weapons Plutonium: Reactor-Related Options," *National Academy of Sciences*, Washington: National Academy of Sciences Press, 1995.

12   *Yucca Mountain Science and Engineering Report*, U.S. Department of Energy, Office of Civilian Radioactive Waste Management, May 2001, p. xxxiii.

13   T. Boone Pickens, (Interviewed). (4 August 2008). *Larry King Live* (Television broadcast). Atlanta, GA: CNN News.

14   "War in Afghanistan (2001–present)," *Wikipedia* (n.d.) at http:// en.wikipedia.org/wiki/War_in_Afghanistan(2001%E2%80%93 present); Internet accessed 5 August 2008. See also: *Today's debate: Oil Supplies. Nation cheers cheap gas, but oil dependency is troubling*, USA TODAY, Monday, 24 December 2001, A10.

15   Pepe Escobar, *The other Iraqi civil war*, Asia Times, 3 April 2008, at http://www.atimes. com/atimes/Middle_East/JD03Ak01.html; Internet accessed 6 August 2008.

16 "Post 9-11 World" being the world that exists subsequent to the terrorist attacks on the United States that were executed on September 11, 2001, which have changed forever the way people in this country live their lives.

17 Currently, Pakistan and India are mobilized along their common border in a nuclear standoff. And rogue nations, such as Iraq, Iran, and North Korea, are particularly worrisome.

18 Pratap Chakravarty, "Missiles In 'Position' As India Mulls Further Action Against Pakistan," *Space Daily*, New Delhi, India, 26 December 2001, at http://www.spacedaily. com/news/nuclear-india-pakistan-01a. html; Internet accessed 10 February 2003.

19 Carey Sublette, *Pakistan's Nuclear Weapons Program: 1998:The Year of Testing*, 10 September 2001, at http://nuclearweaponarchive. org/Pakistan/PakTests.html; Internet accessed 6 August 2008.

20 Walter Pincus, "Report Finds 'Weakness In Nuclear Controls,'" *The Washington Post*, Tuesday, 6 November 2001; A7.

21 Reported by Vivienne Walt, *Uranium reportedly found in tunnel complex*, USA TODAY, 24 December 2001, A6.

22 Walter Pincus, "Nuclear Expert's Nightmare: Terrorists Steal a Warhead," *Washington Post*, 4 November 2001, A6.

23 "Poisoned Spy Blames Putin For His Death: Radioactive Traces Found In Ex-Spy's Home, Haunts," *CBS News Interactive: About Russia*, London, 24 November 2006, at http://wcbstv.com/topstories/Russia. spy.Alexander.2.275580.html; Internet accessed 6 August 2008.

24 Paul L. Williams, *Ramadan and nuke terror: Is 'American Hiroshima' set for this month?*, World Net Daily, 7 October 2005, at http://www. worldnetdaily.com/news/ article.asp?ARTICLE_ID=46705; Internet accessed 6 August 2008.

25 Ewen MacAskill and Chris McGreal, *Israel should be wiped off map, says Iran's president*, The Guardian, London, 27 October 2005, at http://www.guardian.co.uk/ world/2005/oct/27/israel.iran; Internet accessed 6 August 2008.

26   "Islamic extremists invade U.S., join sleeper cells," *The Washington Times,* Washington, DC, 9 February 2004, at http://www.washingtontimes.com/news/2004/feb/09/20040209-115406-6221r/; Internet accessed 6 August 2008.

27   Cheryl L. Smart (Lt. Col., U.S.A.), *THE GLOBAL WAR ON TERROR: MISTAKING IDEOLOGY AS THE CENTER OF GRAVITY* (Issue Paper, Center for Strategic Leadership, U.S. Army War College, July 2005, Volume 08-05), at http://www.carlisle.army.mil/usacsl/Publications/08-05-GWOT.pdf; Internet accessed 9 August 2008.

28   "Homeland Security Director on Potential Terrorist Attacks," *US Department of State International Information Programs Washington File*, 3 December 2001, at http://cryptome.org/strike3.htm; Internet accessed 10 August 2008.

29   Guy Gugliotta, "Technology of 'Dirty Bomb' Simple, but Not the Execution," *Washington Post*, 5 December 2001, A12.

30   Bartlett's, 166.1

31   Richard Rhodes, *The Making of the Atomic Bomb* (New York: Simon and Schuster, 1986) 697 and 773. Edward Teller was a senior research fellow at the Hoover Institution in Stanford, California.

32   Will McNamara, "Is Nuclear Energy on the Verge of a Resurgence?" *PBMR* (Pebble Bed Modular Reactor), Issue Alert (10 January 2001): 1-3.

33   "The Nuclear Green Revolution," May 2008, at http://nucleargreen.blogspot.com/ 2008_05_ 01_archive.html; Internet accessed 10 August 2008.

34   Ibid.

35   Marjorie Mazel Hecht, Nuclear Report, Spring 2001, "The New Nuclear Power," 49-62.

36   The Club of Rome's international secretariat is located in Winterthur (Canton Zurich), Switzerland. Comprised of professionals from the fields of diplomacy, industry, academia and civil society, its

focus is long-term concerns regarding unlimited resource consumption in an increasingly interdependent world. http://www.clubofrome.org/; Internet accessed 19 July 2008.

37 "U.S. Nuclear Power Plants Set Record Highs for Electricity Production, Efficiency in 2007," *Reuters*, Press Release, Wednesday, 6 February 2008. http://www.reuters.com/ article/pressRelease/ idUS205768+06-Feb-2008+BW20080206; Internet: accessed 8 July 2008.

38 "Nuclear Power in America: Today Tomorrow," (Washington, DC, Department of Energy, Office of Nuclear Energy, [n.d.]). http:// www.ne.doe.gov/pdfFiles/ne PrimerWeb1.pdf; Internet accessed 23 July 2008.

39 "NEI Congratulates NRG on Filing Complete License Application for New Texas Reactors," *Nuclear Energy Institute,* 25 September 2007, at http: //www.nei.org/ newsandevents/newsreleases/ neicongratulatesnrg/; Internet accessed 10 August 2008.

40 Matthew Mosk, "Train Wreck Stirs Fear Over Nuclear Freight," *The Washington Post,* Sunday, 20 July 2001, 1.

41 Stuart F. Brown, "How do you feel about nuclear power now?" *Fortune,* 4 March 2002, 130-134.

42 "Achieving International Consensus on the Future of Nuclear Energy," *World Nuclear Association,* London (n.d.), 5, at http://www. world-nuclear.org/sgspeeches/cordoba2000. htm; Internet accessed 13 March 2002.

43 In March of 2001, Senator Pete Domenici introduced the Nuclear Energy Assurance Act of 2001, a bipartisan bill encouraging greater use of nuclear power. It would allocate $406 million for research and development, capital improvements for increasing capacity, restructuring of regulatory and government agencies, and studies on the feasibility of new construction. "News & Events," *Nuclear Energy Institute,* 31 March 2001. http://www.nei.org/newsandevents/ speechesandtestimony/2001/yelvertonkeynoteextended; Internet accessed 8 July 2008.

44   Carla Anne Robbins and Alan Cullison, *Closed Doors: In Russia, Securing Its Nuclear Arsenal Is an Uphill Battle: Despite U.S.Help, Program Faces Resistance, Delays Amid Chill in Relations — A Warehouse Sits Empty,* Wall Street Journal, 26 September 2005.   http://www.sgpproject.org/Personal%20Use%20Only/ClosedDoors09.26.05.htm; Internet accessed 9 August 2008.

45   David B. Ottaway, Robert G. Kaiser, and Madonna Lebling, "After Sept. 11, Severe Tests Loom for Relationship," *The Washington Post,* 12 February 2002, A1.

46   James Woolsey, on "Tony Snow," Fox Channel, 21 October 2001, at 12:12 a.m.

47   "Crude Oil and Total Petroleum Imports Top 15 Countries" (Washington, DC: Energy Information Administration, 30 June 2008). http://www.eia.doe.gov/pub/oil_gas/   petroleum/data_publications/company_level_imports/current/import.html; Internet accessed 9 July 2008

48   Ibid.   Percentage calculated from EIA data.

49   "Seven   Revolutions: Revolution   2:   Resources: Energy" (Washington, DC: Center for Strategic and International Studies, 2006).   http://7revs.csis.org/pdf/resource.pdf; Internet accessed 10 July 2008.

50   Michael Klare, *Bush-Cheney Energy Strategy: Procuring the Rest of the World's Oil,* FPIF-PetroPolitics Special Report, January 2004

51   "Oil climbs past $147 on tensions in Iran, Nigeria," *cbcnews.ca,* 11 July 2008. http://www.cbc.ca/money/story/2008/07/11/oilrecord.html; Internet accessed 19 July 2008.

52   "Fact Sheet: Hydrogen Fuel: a Clean and Secure Energy Future" (Washington, DC, The White House, 6 February 2003). http://www.whitehouse.gov/news/releases/2003/02/20030206-2.html;   Internet accessed 19 July 2008.

53   James A. Lake, "Outdated Thinking Is Holding us Back," *The Washington Post,* Sunday, 13 May 2001, B3.

54 Yucca Mountain is a geologic repository or disposal site; it is not a long-term storage site. With any kind of reactor and reprocessing system, a disposal site is *always* necessary. *Interfax* News, 29 April 2001.

55 "Yucca Mountain Project: Information on Project Costs" (Washington DC: Government Accountability Office ltr. to The Honorable Jon C. Porter House of Representatives; Subject: *Yucca Mountain Project: Information on Estimated Costs to Respond to Employee E-mails That Raised Questions about Quality Assurance* (19 January 2007)

56 "Waste Management in the Nuclear Fuel Cycle," *World Nuclear Association,* London, April 2007. http://www.world-nuclear.org/info/inf04.html; Internet accessed 10 July 2008

57 *Washington Post,* 13 May 2001, B3; "The Economics of Nuclear Power," Nuclear Issues Briefing Paper # 8, *Uranium Information Centre* (Melbourne, Australia: July 2001), 1-3, at http://www.uic.com.au/nip08.htm; Internet accessed 2 February 2002. See also: James A. Lake, "Outdated Thinking Is Holding Us Back," *The Washington Post,* Sunday, 13 May 2001, B3.

58 "Nuclear Waste Disposal" (Nuclear Energy Institute, Washington, DC, 18 April, 2008). http://www.nei.org/resourcesandstats/documentlibrary/nuclearwastedisposal/; Internet accessed 13 July 2008.

59 Matthew Mosk, "Train Wreck Stirs Fear Over Nuclear Freight, *The Washington Post,* Sunday, 29 July 2001, A1. http://www.nukewatch.org/media/more_media/08-00-01/twsfonf.html; Internet accessed 10 July 2008. See also: Rebecca Ullrich, "The History of Transportation Technology Programs at Sandia," (a paper prepared in 1995 and last modified 1 October 2001); p. 2. http://www.sandia.gov/recordsmgmt/ctb.html; Internet accessed 25 September 2002.

60 "Disposition of Surplus Highly Enriched Uranium Final Environmental Impact Statement," *U.S. Environmental Protection Agency: Federal Register Environmental Documents* [Federal Register: June 28, 1996 (Volume 61, Number 126)], at http://www.epa.gov/

EPA-IMPACT/1996/June/Day-29/pr-16692.html; Internet accessed 11 August 2008.

61   "Environmental Impact Statement for a Geologic Repository for the Disposal of Spent Nuclear Fuel and High-Level Radioactive Waste at Yucca Mountain, Nye County, Nevada" (Washington, DC: U.S. DOE Office of Civilian Radioactive Waste Management, vol. 1, July 1999), chaps. 1-7.

62   "Adoption of Final Environmental Impact Statement," *U.S. Environmental Protection Agency: Federal Register Environmental Documents* [Federal Register: February 14, 2001 (Volume 66, Number 31)], at http://www.epa.gov/EPA-IMPACT/2001/February/ Day-14/ i3693.htm; Internet accessed 11 August 2008.

63   *(U.S.) Central Intelligence Agency: World Factbook, at http://* www.cs.fit.edu/~ryan/factbook/factbook2007/geos/xx.html; Internet accessed 11 August 2008.

64   Department of Energy, Environmental Impact Assessment Reports, July 1999, DOE/EIS-0250D, at http://www.eia.doe.gov/ oiaf/archive/aeo01/overview.html; Internet accessed 2 February 2002.

65   "Analysis of Increased Power Costs in California: Why California Power Costs Increased in 2000 and Who Collected the Revenues," *Reliant Energy,* 22 June 2001, 1-13.

66   "California's Electricity," *World Nuclear Association* London, July 2008, at http://www.world-nuclear.org/info/inf65.html; Internet accessed 11 August 2008.

67   "Blackout of 2003: How Did It Happen and Why?" (Text of Full Committee on Energy and Commerce), *Global Security* (4 September 2003), at http://www.globalsecurity.org/ security/library/ congress/2003_h/030904-house-energy-commerce_transcript.htm; Internet accessed 11 August 2008.

68   "The Western Energy Crisis, the Enron Bankruptcy, and FERC's Response," *Federal Energy Regulatory Commission* (n.d.), at http://

www.ferc.gov/industries/electric/indus-act/wec/chron/chronology.pdf;
Internet accessed 11 August 2008.

69   "The Outlook for Nuclear Energy in a Competitive Electricity
Business," Fact Sheet, *Nuclear Energy Institute* (2002), 1-10, at http://
www.nei.org/doc.asp?Print=true&DocID=&CatNum+f3&CatI
D=38; Internet accessed 10 March 2002. See also: http://www.nei.
org/resourcesandstats/nuclear_statistics/usnuclearpowerplants/, which
reports nuclear power plants operated at 91.8 % for 2007.

70   "Nuclear Power in the USA," *World Nuclear* Association, London,
August 2008, at http://www.world-nuclear.org/info/inf41.html;
Internet accessed 11 August 2008.

71   Since Tauzin's remarks, the NRC has approved renewal for 43
reactors; total extensions to date is 48.   "Backgrounder on Reactor
License Renewal" (Nuclear Regulatory Commission, Washington,
DC, April 2007). http://www.nrc.gov/reading-rm/doc-collections/
fact-sheets/license-renewal-bg.html; Internet accessed 13 July 2008.

72   Bill # S.1653, introduced by James Inhofe on 19 June 2007,
proposes "(A bill) to implement the Convention on Supplementary
Compensation for Nuclear Damage, and for other purposes." Price-
Anderson is being kept alive by the Congress. http://www.congress.
org/congressorg/issues/bills/?billnum=S.1653&congress=110; Internet
accessed 13 July 2008.

73   "Abraham Announces Plan to Dispose of Surplus Plutonium,"
Federation of American Scientists, Washington, DC (24 January 2002),
at http://www.fas.org/news/usa/2002/012402was.htm; Internet accessed
11 August 2008.

74   "Mixed Oxide Fuel Fabrication Facility Licensing," *U.S. Nuclear
Regulatory Commission* (n.d.), at http://www.nrc.gov/materials/fuel-
cycle-fac/mox/licensing.html; Internet accessed 11 August 2008.

75   Guy Gugliotta, *Nuclear Reprocessing Sets Off Alarms Again:
Comment in Bush Plan Re-Energizes Old Debate,* Washington Post,
Washington, DC (2 July 2001). Available on-line at http://nucnews.

net/nucnews/2001nn/0107nn/010702nn.htm; Internet accessed 11 August 2008.

76 Anthony Andrews, *Nuclear Fuel Reprocessing: U.S. Policy Development,* Congressional Research Service Report for Congress (Order Code RS22542, Updated March 27, 2008}, at http://www.fas.org/sgp/crs/nuke/RS22542.pdf; Internet accessed 11 August 2008.

77 Mel Buckner, reporting on the House/Senate Nuclear Caucus event on March 5 and 6, 2001; the event was entitled "U.S. Leadership," "MOX: SWORDS INTO PLOWSHARES" (Position Paper), *CNTA Speakers' Bureau,* Aiken, SC, Spring 2001, at http://cache.search.yahoo.net/search/cache?ei=UTF-8&p=mel+buckner+us+leadership+ house+senate+nuclear+caucus+march+2001&fr=yfp-t-501&u=www.c-n-t-a.com/awaremain_files/aware0103.pdf&w=mel+buckner+us+leadership+leader+leaders+%22leader+ship%22+house+senate+nuclear+caucus+march+2001&d=Uy3LMhg5RQYp&icp=1&.intl=us; Internet accessed 11 August 2008.

78 Joseph Curl, "Bush Sees Dual Goals of Energy, Environment," *Washington Times,* 18 May 2001.

79 Juliet Eilperin and Eric Pianin, *Washington Post,* 1 August 2001, A2.

80 Mike Allen and Eric Pianin, *Washington Post,* 9 May 2001, A14.

81 Ibid.

82 "Advanced Fuel Cycle Initiative," *U.S. Department of Energy* (n.d.), at http://www.ne.doe.gov/AFCI/neAFCI.html; Internet accessed 11 August 2008.

83 See: Statement of Paul M. Golan, Acting Director for the Office of Civilian Radioactive Waste Management, U.S. Department of Energy, Before the Subcommittee on Energy and Water Development, and Related Agencies, U. S. House of Representatives (March 15, 2006), at http://www.ocrwm.doe.gov/info_library/program_ docs/speeches/hymp-progmgmt-fy07-budget.pdf; Internet accessed 11 August 2008.

84 President George W. Bush (Kwame Holman [Online News Hour] outlines President Bush's energy plan and the initial reactions. St. Paul, Minnesota, 17 May 2001) http://www.google.com/search?hl=en&q=T he+truth+is+that+energy+production+and+environmental+protection +are+not+competing+priorities.++&btnG=Google+Search

85 "Carbon's Burden on the World's Oceans, *Environment 360* (n.d.), at http://e360.yale.edu/content/feature.msp?id=1996; Internet accessed 11 August 2008.

86 "Nuclear Power: Nuclear Disarmament," *Greenpeace USA* (n.d.), at http://www.greenpeaceusa.org/nuclear/nuclearpowertext.htm; Internet accessed 4 April 2002.

87 "Greenpeace: About us," (n.d.) http://www.greenpeace.org/ aboutus; Internet accessed 25 September 2002.

88 Patrick Moore, "Going Nuclear: A Green Makes the Case," *The Washington Post*, Sunday, April 16, 2006; B01. http://www. washingtonpost.com/wpdyn/content/article/2006/04/14/ AR2006041401209.html; Internet accessed 20 July2008.

89 "France and the fight against global warming," *France in Australia Embassy and Consulate-General* (n.d.), at http://www.ambafrance-au. org/article.php3?id_article=1937; Internet accessed 11 August 2008.

90 Bartlett's, 73.6.

91 Madlen Read, "Oil Tops $147 Setting New Record," *New York Sun*, Friday, 11 July 2008. http://www.nysun.com/business/oil-sets-new- record-near-147-a-barrel/81703/; Internet accessed 17 July 2008.

92 *The American Heritage College Dictionary* (1993), 3rd. ed., s.v. "vitrification."

93 The distinction between "rogue-led states" and "rogue states" was made by Austin Bay in "Serious budget for a serious war," *The Washington Times*, Friday, 12 February 2002, A21.

94 "President Delivers State of the Union Address," *The White House*, Washington, DC (29 January 2002), at http://www.google.com/sear

ch?hl=en&q=+Bush+said+in+his+ first+State+of+the+Union+address; Internet accessed 11 August 2008.

95 "RADIOLOGICAL DISPERSAL DEVICES: AN INITIAL STUDY TO IDENTIFY RADIOACTIVE MATERIALS OF GREATEST CONCERN AND APPROACHES TO THEIR TRACKING, TAGGING, AND DISPOSITION" (Report to the Nuclear Regulatory Commission and the Secretary of Energy, May 2003), at http://www.nti.org/ e_research/official_docs/doe/ DOE052003.pdf; Internet accessed 11 August 2008.

96 Walter Pincus, "Report Finds 'Weakness In Nuclear Controls,'" *The Washington Post,* Tuesday, 6 November 2001; Pg. A07.

97 "The World's Nuclear Arsenals," *Center for Defense Information,* Washington, DC (Last updated 4 February 20003), at http://www. cdi.org/issues/nukef&f/database/ nukearsenals.cfm; Internet accessed 11 August 2008.

98 Haizam Amirah-Fernández, *Will the Fourth Gulf War be Avoided?,* Real Instituto Elcano, Madrid, Spain (7 December 2007), at http:// www.realinstitutoelcano.org/wps/portal/riel cano_in/Content?WCM_ GLOBAL_CONTEXT=/Elcano_in/Zonas_in/ARI+62-2007; Internet accessed 11 August 2008.

99 Susan A. Kitchens, *An Oppenheimer discussion at Cal Tech,* 2020 Hindsight (2 February 2006).

100 Rhodes, 756.

101 Kai Bird and Sherwin Martin, "The First Line Against Terrorism," *Washington Post,* 12 December 2001, A35.

102 "U.N. nuke team searches Georgia woods for containers," *Reuters, Nuclear News* (11 June 2002), at http://nucnews.net/ nucnews/2002nn/0206nn/020611nn.htm#010; Internet accessed 12August 2008.

103 Richard Morino and Claudia Dean, "Rethinking Coverage on the Front Lines," *Washington Post,* 8 January 2002, A15. Also visit the NTI home page at www.nti.org.

104   Karen De Young and Dana Milbank, "Bush, Putin Agree to Slash Nuclear Arms," *Washington Post,* 14 November 2001, A1.

105   Mike Allen and Philip P. Pan, "Bush and Putin Edge Closer to Missile Deal, *Washington Post,* 22 October 2001, A1.

106   Michael Evans, RAF jets scrambled to intercept Russian bomber," *Times Online: UK Edition,* (22 August 2007), at http://www.havertys. com/View_dogDays/index?cm_mmc=   casale-_-promo2dogdays-_- null-_-null; Internet accessed 12 August 2008.

107   "Chart: Al-Qa`ida's WMD Activities , *Center for Nonproliferation Studies (CNS)* (Updated 13 May 2005), at http://cns.miis.edu/pubs/ other/sjm_cht.htm; Internet accessed 12 August 2008.

108   Mission and Goals of the Missile Defense Agency (MDA) may be found on its Home Page, at http://www.winmda.com/; Internet accessed 12 August 2008.

109   "N. Korea Said Trying to Sell Missiles," *New York Times*, Friday, 9 November 2001.

http://www.prop1.org/nucnews/2001nn/0111/011109nn.htm; Internet accessed 13 March 2002.

110   "Bush Labels North Korea, Iran, Iraq an 'Axis of Evil'," *Arms Control Association,* Washington, DC (March 2002), at http://www. armscontrol.org/contact; Internet accessed 12 August 2008.

111   Walter Pincus, "Nuclear Strike on Bunkers Assessed," *Washington Post*, 6 November 2001, A28

112   Karen DeYoung and Dana Milbank, "U.S. Repeats Warnings on Terrorism: Bush Urges Other Nations to 'Get Their House in Order,' " *Washington Post*, 1 February 2002, A1.

113   Letter to Joseph Milligan, April 6, 1816, *The Proceedings of the Friesian School: Fourth Series*, at http://www.friesian.com/; Internet accessed 24 August 2008.

114   "World Uranium Mining," *World Nuclear Association*, London, July 2008, at http://www.world-nuclear.org/info/inf23.html; Internet accessed 12 August 2008.

115   Ibid.

116   "NRC ISSUES LICENSE TO LOUISIANA ENERGY SERVICES FOR GAS CENTRIFUGE URANIUM ENRICHMENT PLANT IN NEW MEXICO," *U.S. NUCLEAR REGULATORY COMMISSION*, Washington, DC   (23 June 2006), at http://adamswebsearch2.nrc.gov/idmws/doccontent.dll?library=PU_ADAMS^PBNTAD01&ID=061770101; Internet accessed 13 August 2008.

117   Phillip J. Finck, *Argonne National Laboratory,* (Statement) Before the House Committee on Science, Energy Subcommittee Hearing on Nuclear Fuel Reprocessing, Washington, DC   (16 June 2005), at http://www.anl.gov/Media_Center/News/2005/   testimony050616.html; Internet accessed 13 August 2008.

118   Ruth Lewin Sime, *Lise Meitner: A Life in Physics* (University of California Press, 1996), 305.

119   "In Some Physics of Uranium," *World Nuclear Association* (Information Paper) London (n.d.), at http://www. world-nuclear.org/education/phys.htm; Internet accessed 11 September 2008. Visit this website for a more detailed explanation of the nuclear fission process.

120   "Lise Meitner: A Life in Physics," 305.

121   "Russia's GA to Burn Weapons Plutonium," 34.

122   *The American Heritage College Dictionary, Third Edition* (Houghton Mifflin Company: Chicago, 1992).

123   "Why Uranium," *Trigon Uranium Corp.,* Golden, Colorado, 2008. http://www.trigonuraniumcorp.com/why_uranium/; Internet accessed 21 July 2008.

124   Rhodes, 403 and 476.

125   "Dual Capable Nuclear Technology," *Large and Associates* (A report prepared for Greenpeace U.K., Ref. No. LARL 2084-A), at http://archive.greenpeace.org/comms/nukes/nukes.html;   Internet accessed 9 August 2002.

126   "Technology heralds the greening of nuclear energy," News, *The University of Sydney* (8 June 2006) at http://www.usyd.edu.au/news/84.html?newsstoryid=1095; Internet accessed 30 August 2008.

127   [127] Peter G. Tsouras, "The Book of Military Quotations: Cicero Ad Atticum: Themistocles (c. 528-462 BC)," http://books.google.com/books?id=BDgnD0omDo0C&pg=PA395&lpg= PA395&dq= C icero+Ad+Atticum+themistocles,+he+who+commands&source=web &ots= 6pabC1g8X3&sig=gMwqLC1eobJ3a_JvvcnTtM4PyoY&hl=e n&sa=X&oi=book_result&resnum=4&ct=result; Internet accessed 24 August 2008.

128   "MILNET: Naval   War   Ships."   http://www.milnet.com/navywars.htm; Internet accessed 9 August 2008.

129   "Submarines: Underway on Nuclear Power," *National Geographic: Xpeditions,* (n.d.), at http://www.nationalgeographic.com/xpeditions/lessons/13/g912/k19underway. Htm; Internet accessed 13 August 2008.

130   "History of Nuclear Power in the Navy," *U.S. Navy,* at http://www.cnrc.navy.mil/nucfield/Background/history.htm; Internet accessed 8 August 2008.

131   "Key Benefits of Utilizing Nuclear Energy to Power Ships," *U.S. Navy.*

http://www.cnrc.navy.mil/nucfield/Background/why.htm;   Internet accessed 8 August 2008.

132   "Naval Reactors," *U.S. Navy,* at *http*://www.cnrc.navy.mil/nucfield/Background/ naval.htm; Internet accessed 10 August 2008.

133   "United States Atomic Energy Commission," *Wikipedia, the free encyclopedia, at http*://en.wikipedia.org/wiki/United_States_Atomic_Energy_Commission; Internet accessed 10 August 2008.

134 (Untitled article about Captain Hyman G. Rickover, U.S. Navy, and Establishment of the Naval Nuclear Propulsion Program, at http://www.cnrc.navy.mil/nucfield/Background/ naval.htm; Internet accessed 10 August 2008.

135 Ibid.

136 Ibid.

137 Ibid.

138 "Operation Sea Orbit," *U.S. Navy,* at *http*://www.cnrc.navy.mil/ nucfield/ Background/enterprise.htm; Internet accessed 10 August 2008.

139 Admiral Hyman G. Rickover, *Doing a Job,*(excerpts from a speech delivered at Columbia University in 1982) GovLeaders.org, at http://www.govleaders.org/rickover.htm; Internet accessed 10 August 2008.

140 Knolls Atomic Power Laboratory (KAPL), at http://www. kaplinc.com/index.html; Internet accessed 10 August 2008.

141 "History," *Idaho National Laboratory,* at *http*://www.inl.gov/ history/; Internet accessed 10 August 2008.

142 "Cores and Competencies," *Idaho National Laboratory,* at *http*:// www.inl.gov/proving-the-principle/chapter_10.pdf; Internet accessed 10 August 2008.

143 "Advanced Test Reactor National Scientific User Facility," *Idaho National Laboratory,* at *http*s://inlportal.inl.gov/portal/server.pt?open= 514&objID=1360&parentname= CommunityPage&parentid=9&m ode=2&in_hi_userid=200&cached=true; Internet accessed 10 August 2008.

144 "Contamination at INL: Text of the 1995 Settlement Agreement," *Idaho Department of Environmental Quality,* Boise, Idaho, 16 October 1995, at http://www.deq.state.id.us/ inl_oversight/contamination/ settlement_agreement_.cfm; Internet accessed 10 August 2008.

145   See other testimonials at: http://www.cnrc.navy.mil/nucfield/ Background/ testimonials.htm; Internet accessed 10 August 2008.

146   "Trouble at a Reactor? Call In an Admiral," *The New York Times: Business,* 10 August 2008, at http://query.nytimes.com/gst/fullpage.h tml?res=950DEED61E38F934A25751C0 A96F948260&sec=&spon =&pagewanted=all; Internet accessed 10 August 2008.

147   "Nuclear Propulsion," *Federation of American Scientists: Military Analysis Network,* (n.d.).   http://www.fas.org/man/dod-101/sys/ship/ eng/reactor.html; Internet accessed 23 July 2008.

148   Ibid.

149   Ibid.

150   Ibid.

151   Robert O. Work, "THE 313-SHIP FLEET AND THE NAVY'S 30-YEAR SHIPBUILDING PLAN: AFFORDABILITY AND ISSUES" (Congressional Testimony), *Center for Strategic and Budgetary Assessments* (30 March 2006) at http://www.globalsecurity. org/military/library/congress/2006_hr/060330-work.pdf;   Internet accessed 31 August 2008.

152   Ibid.

153   Ibid.

154   "Concern for the Environment," *National Nuclear Security Administration,* U.S. Department of Energy, Washington, DC (n.d.), at http://nnsa.energy.gov/naval_ reactors/print/concern_environment. htm; Internet accessed 13 August 2008.

155   U.S. Environmental Protection Agency: National Priorities List (NPL), at http://oaspub.epa.gov/oerrpage/advquery; Internet accessed 17 August 2008.

156   President William J. Clinton as quoted in "Over 117 Million Miles Safely Steamed on Nuclear Power: The United States Naval Nuclear Propulsion Program" (Washington, DC: U.S. Department of Energy and U.S. Department of Defense, August 1999).

157   Tributes by the Congress, three presidents, et.al. may be viewed online: http://www.cnrc. navy.mil/nucfield/Background/testimonials. htm; Internet accessed 13 August 2008.

158   Ibid.

159   Ibid.

160 "History of USS *Thresher* (SSN-593)," *Naval Historical Center* (Washington, DC, n.d.) at http://www.history.navy.mil/danfs/t/ thresher.htm; Internet accessed 27 September 2008.

161   Online, *Molten Eagle* (31 May 2007), at http://aquilinefocus. blogspot.com/ 2007_05_01_archive.html; Internet accessed 12 August 2008.

162   "SUBMARINE CLASSES," *MILNET* (n.d.), at http://www. milnet.com/pentagon/ subclass.htm#subclasses; Internet accessed 12 August 2008.

163   Ibid.

164   "The Christening," George H. W. Bush (CVN 77) Website - Northrop Grumman, Newport News, Virginia, at http://www. nn.northropgrumman.com/bush/christening.html; Internet accessed 13 August 2008. See CVN 77 Commissioning; Internet 20 August 2009.

165   Henry Stephens Randall, "The Life of Thomas Jefferson," Letter to Maria Jefferson Eppes, 26 October 1801, 675, at http:// books.google.com/books?id=lxxCAAAAIAAJ &pg= PA675&lpg=P A675&dq=thomas+jefferson+%22a+little+experience+is+worth+ a +great+deal+of+reading%22&source=web&ots=YUxu9NPhIg&sig =0Eu6SHq7QXC7WZTfcnkQrP7buog&hl=en&sa=X&oi=book_ result&resnum=3&ct=result; Internet accessed 24 August 2008.

166   John Ardagh, *FRANCE in the New Century: Portrait of a Changing Society* (London: Penguin Books, 2000), 1-757.

167   Ibid., 115.

168   Ibid., 115-116.

169   Ibid., 116-117.

170   Ibid., 117.

171   "How Do You Feel About Nuclear Power Now?" *Fortune*, 4 March 2002, 130-134.

172   "Nuclear Power in France" (n.d.), at http://www.info-france-usa.org/ america/ embassy/ nuclear/profile/introduc/industry.htm; Internet accessed 21 March 2002.

173   "New research reveals the real cost of electricity in Europe," *Research* (Brussels, 20 July 2001), at http://europa.eu.int/comm/ research/press/2001/pr2007en.html; Internet accessed 10 February 2003.

174   Ardagh, 117.

175   " Nuclear Power in Italy," *World Nuclear* Association, London, June 2008. http://www.world- nuclear.org/info/inf101.html; Internet accessed 20 July 2008.

176   Elizabeth Rosenthal, "Italy Plans to Resume Building Atomic Plants," *New York Times*, 23 May 2008. http://www. nytimes.com/2008/05/23/world/europe/   23nukes.html? partner=rssnyt&emc=rss; Internet accessed 20 July 2008.

177   "Flamanville Nuclear Power Plant," *Wikipedia, at http://* en.wikipedia.org/wiki/ Flamanville_Nuclear_Power_Plant; InSternet accessed 12 August 2008.

178   "Nuclear Power in France- Why does it work?" http://www. npcil.nic.in/nupower_vol13_2/npfr_.htm; Internet accessed 13 August 2008.

179   "Nuclear Power in France," *World Nuclear Association*, London, May 2008.   http://www.world-nuclear.org/info/inf40.html; Internet accessed 20 July 2008.

180   Bartlett's, 359.8.

181   Paul E. Gray," Nuclear Reactors Everyone Will Love," *The Wall Street Journal*, Thursday, 17 August 1989.

182   Martin I. Hoffert, "Technologies for a Greenhouse Planet," *Department of Physics, New York University*, New York (n.d.) at http://physics.nyu.edu/preprint/hoffert. martin.99A.pdf; Internet accessed 24 August 2008.

183   "Advanced Reactors," *World Nuclear Association*, London, November 2002, at http:www.world-nuclear.org/info/inf08.htm, 2; Internet accessed 21 January 2003.

184   "How the Pebble Bed Modular Reactor Works," PBMR (n.d.), 1-4, at http://www.pbmr.com/2_about_the_pbmr/2_8background_to_the_pbmr.htm; Internet accessed 13 April 2002.

185   Dave Nicholls, "The Pebble Bed Modular Reactor," *Nuclear News*, 2 September 2001, 35.

186   "Constellation Energy and EDF Form Joint Venture for Developing Next-Generation Nuclear Facilities in the United States and Canada," (n.d.). http://www.prnewswire.com/ cgi-bin/stories.pl?ACCT=104&STORY=/www/story/07-20-2007/ 0004629591& EDATE=; Internet accessed 20 September 2008.

187   "Status of Potential New Commercial Nuclear Reactors in the United States," *Energy Information Administration,*17 July 2008. http://www.eia.doe.gov/cneaf/nuclear/ page/nuc_reactors/reactorcom.html; Internet accessed 21 July 2008.

188   Gas-Cooled, 'Passive,' Medium-Sized: Gas Turbine Modular Helium Reactor (GT-MHR)," *Nuclear Energy Institute* (2002), 1-2. http://www.nei.org/doc.aspp?Print=true&DocID=&CatNum=3&CatID=711Internet accessed 21 March 2002. 2); "GT-MHR: Inherently Safe Nuclear Power for the 21st Century," *General Atomics* (2002), 1, at http://www.gat.com/gtmhr; Internet accessed 21 March 2002.

189   Marjorie Mazel Hecht, "A Meltdown Proof Reactor. The General Atomics GTR: The New Nuclear Power," Nuclear Report, *21st Century Science and Technology* (Spring 2001), 55.

190   Alan Wong, "The Fifty Percent Efficiency Nuclear Power Plant," Department of Nuclear Engineering, University of California, Berkeley, NE-161 Project Fall 1994 (n.d.), at http://www.nuc.berkeley.edu/thyd/ne161/alwong/ne161.html; Internet accessed 5 April 2002; "GT-MHR: Inherently Safe Nuclear Power for the 21st Century," General Atomics (2002): 1, at http://www.gat.com/gtmhr; Internet accessed 21 March 2002.

191   "Russia's GA to Burn Weapons Plutonium," 34.

192   For an informative discussion of gas turbine technology, including nuclear, see: Lee S. Langston, "Power Energy: Wild Blue Yonder," *Mechanical Engineering* (May 2006) at http://www.memagazine.org/backissues/membersonly/may06/features/wildblue/wildblue.html; Internet accessed 31 August 2008. See also, end note 190.

193   T. A. Donohue, General Manager, Advanced Technology Operations, General Electric, Dr. J. A. Friedericy, Director, Research & Technology, Allied Signal Aerospace. "Russia's GA to Burn Weapons Plutonium," 34.

194   "The International Nuclear Event Scale (INES)" at http://www.atomeromu.hu/ biztonsagINESskala-e.htm; Internet accessed 28 September 2008.

195   V. Galushkin, State Unitary Company, Experimental Design Bureau for Mechanical Engineering (OKBN), *Health Physics & Waste & Fuel Management: GT-MHR—An Advanced Reactor,* Nuclear Plant Journal, vol. 19, no.1, 39.

196   Ibid.

197   Paul E. Gray, "Nuclear Reactors Everyone Will Love." *The Wall Street Journal,* Thursday, August 17, 1989, A-1.

198   Bartlett's, 99.12.

199   "Waste Management in the Nuclear Fuel Cycle," Nuclear Issues Briefing Paper 9, *Uranium Information Centre* (September 2001), 3. http://www.uic.com.au/nip09.htm; Internet accessed 21 March 2002.

200   Pradel. c1998.

201   Ph. Pradel et.al., "Waste Minimization and Management Along Processing and Recycling of Nuclear Materials," Abstract Joint Cogema and SGN (c1998). http:// www.wmsym.org/wm98/htmq/sess08/08-02.htm; Internet accessed 2 February 2002.

202   "Radioactive Wastes: International organizations and safety standards," *World Nuclear Association,* London, March 2001, at http:// www.worldnuclear.org/info/ inf60.html; Internet accessed 17 August 2008.

203   "Reprocessing," (n.d.), 1-2, at http://www.nuke-energy.com/ data/reprocessing. html# france; Internet accessed 18 March 2002.

204   "Status and Advances in MOX Fuel Technology" (Technical Reports Series No. 415) *International Atomic Energy Agency* (Vienna 2003) at http://www-pub.iaea.org/MTCD publications/PDF/ TRS415_web.pdf; Internet accessed 30 September 2008.

205   "Management and Disposition of Excess Weapons Plutonium," 113

206   "Environmental Impact Statement for a Geologic Repository for the Disposal of Spent Nuclear Fuel and High-Level Radioactive Waste at Yucca Mountain, Nye County, Nevada" (Washington, DC: U.S. DOE Office of Civilian Radioactive Waste Management, vol. 1, July 1999), chaps. 1-7.

207   In the vitrification process, chunks of frit are combined with waste, melted, and then cooled to form a vitrified product. See: "WASTE VITRIFICATION: MORE THAN ONE STRING TO ITS BOW," *CEA* (2002) at http://www.cea.fr/var/cea/storage/static/ gb/library/Clefs46/pagesg/cleInternet accessed 21 August 2008 .

208   Thomas Jefferson, *The Writings of Thomas Jefferson,* 1903, 311, Original from Harvard University, Digitized Dec 5, 2007 at http:// books.google.com/books?id= omQSAAAAYAAJ&dq=thomas+jefferso n+%22what+has+passed+may+be+a+lesson%22; Internet accessed 24 August 2008.

209　*News-Globe*, Amarillo, Texas, 3 November 1996, 9B, at http://www.nci.org/a/a.11396.htm; Internet accessed 14 March 2002.

210　Sophocles, *The Quotations Page,* at http://www.quotationspage.com/ quote/24117. html; Internet accessed 24 August 2008.

211　"Using the Delphi Technique to Achieve Consensus," *Education Reporter* (November 1998), at http://www.eagleforum.org/educate/1998/nov98/focus.html.; Internet accessed 21 March 2002.

212　"Reactor-Grade and Weapons-Grade Plutonium in Nuclear Explosives" (Washington, DC: an excerpt from DOE publication, "Nonproliferation and Arms Control Assessment of Weapons-Usable Fissile Material Storage and Excess Plutonium Disposition Alternatives," January 1997), 37-39.

213　"The 'Back-End' of the Nuclear Fuel Cycle," *Nuclear Electricity*, 6th ed., August 2000, updated November 2001, Ch. 5, pp. 3-5.

214　"A Technology Roadmap for Generation IV Nuclear Energy Systems" (Washington, DC, December 2002). http://www.ne.doe.gov/genIV/documents/gen_iv_roadmap.pdf; Internet accessed 23 July 2008.

215　"Technology Goals for Generation IV Nuclear Energy Systems," was approved by the Generation IV Roadmap Nuclear Energy Research Advisory Committee (NERAC) Subcommittee (GRNS) on April 13, 2001, for presentation to the NERAC on May 1, 2001. The DOE report is dated 13 April 2000, at http://gen-iv.ne.doe.gov; Internet accessed 21 March 2002.

216　"The History of Nuclear Power Plant Safety: Advanced Reactor Research in High Gear," Nineties,(c1999), 1-6, at http://users.owt.com/smsrpm/nksafe/nineties.html; Internet accessed 21 March 2002.

217　Donald G. Newman, Jerome P. Lavelle, and Ted G. Eschenbach, *Engineering Economic Analysis,* 8th ed. (Austin, Tex.: Engineering Press, 2000), 606-608.

218   Rebecca Smith, "Industry Argues New Designs Would Be Safe, Cost Less; But Will Public Buy That?" *Wall Street Journal*, 2 May 2001, B1.

219   "Management and Disposition of Excess Weapons Plutonium: Reactor Related Options" (Washington, DC: National Academy of Sciences, National Academy Press, 1995), 401.

220   Alfred D. Chandler, Jr., *Scale and Scope: The Dynamics of Industrial Capitalism* (Cambridge: Harvard University Press, Belknap Press, 1990), 3-46.

221   "Over 117 Million Miles Safely Steamed on Nuclear Power," The United States Naval Nuclear Propulsion Program, U.S. Department of Energy and U.S. Department of Defense (Washington, DC: August 1999). U.S. Navy Nuclear Propulsion Program Information, at http://www.subasekb.navy.mil/propels.htm; Internet accessed 23 March 2002.

222   DOE Executive Safety Conference, December 11-12, 2001, Session 3 "Improving the Contribution of Operating Experience, Performance Monitoring and Analysis, and Lessons Learned to Integrated Safety Management (Feedback for Improvement)" addressed *Performance metrics (How do we know how we are doing?)*, at http://www.tis.eh.   doe.gov/ism/events/workshops/exec2001/final.pdf: Internet accessed 10 February 2003.

223   Joseph A Maxwell, *Qualitative Research Design: An Interactive Approach* (Thousand Oaks, Calif.: Sage Publications, 1996), 95.

224   A decision-making technique developed by the Rand Corp. in the 1960s. See "Prioritization Process Using Delphi Technique" at http://www.carolla.com/wp-delph.htm.

225   Hossein Arsham, "Questionnaire Design and Surveys Sampling," *Europe Mirror,* (n.d.), at http://ubmail.ubalt.edu/~harsham/stat-data/opre330Surveys.htm: Internet accessed 4 February 2003.

226 Sophocles, Antigone, *The quotations Page,* at http://www. quotationspage.com/ quote/24113.html; Internet accessed 24 August 2008.

227 Refer to any textbook on Engineering Economy to learn more about tangible and intangible factors in economic analyses of alternatives. Donald G. Newman, Jerome P. Lavelle, and Ted G. Eschenbach. 2000. *Engineering Economic Analysis,* 8th ed. Austin, Tex.: Engineering Press.

228 "Energy Analysis of Power Systems," Nuclear Issues Briefing Paper # 57, *Uranium Information Centre* (September 2001), chap. 3, 4, at http://www.uic.com.au/nip57.htm; Internet accessed 2 February 2002.

229 Office of Nuclear Reactor Regulation (U.S. Nucear Regulatory Commission, n.d.) at http://www.nrc.gov/about-nrc/organization/ nrrfuncdesc.html#dlr; Internet aaccessed 28 September 2008.

230 Gail Cohen and others, "Electric Utilities: Deregulation and Stranded Costs," *Congressional Budget Office* (October 1998): 1-20.

231 Bruce Biewald and David White. "Stranded Nuclear Waste: Implications of Electric Industry Deregulation for Nuclear Plant Retirements and Funding Decommissioning and Spent Fuel," *Synapse Energy Economics* (January 15, 1999): 1-53.

232 "Costs: Fuel, Operation and Waste Disposal: Operations and Maintenance (O&M) Costs," *Nuclear Energy Institute* (n.d.), at http:// www.nei.org/resourcesandstats/ nuclear_statistics/costs/; Internet accessed 17 August 2008.

233 "Energy Analysis of Power Systems," Nuclear Issues Briefing Paper # 57, *Uranium Information Centre* (September 2001), 2; Internet accessed 8 February 2002.

234 "Energy Analysis of Power Systems," Nuclear Issues Briefing Paper # 57, *Uranium Information Centre* (September 2001), 2; Internet accessed 8 February 2002.

235   "Interview with Linden Blue," *Nuclear Report: 21st Century* (Spring 2001): 57; "Russia's GA Reactor to Burn Weapons Plutonium," *Science and Technology* (March 1, 2002): 34-41.

236   "The Economics of Nuclear Power," Nuclear Issues Briefing Paper # 8, *Uranium Information Centre* (January-February 2000), 2. http://www.uic.com.au/nip08.htm; Internet accessed 2 February 2002.

237   "U.S. Nuclear Power Plants Set Record Highs For Electricity Production, Efficiency in 2007," *U.S. Nuclear Energy*, Washington, DC, 6 February 2008. http://www. usnuclearenergy.org/2007_Plant_Production.htm; Internet accessed 25 July 2008.

238   "The Economics of Nuclear Power," *World Nuclear Association*, London, July 2008, at http://www.world-nuclear.org/about/contact. html; Internet accessed 25 July 2008.

239   "USA: US nuclear cost pips coal," Uranium Information Centre (UIC) Newsletter # 1, *Uranium Information Centre* (January-February 2000), at http://www.uic.com.au/ news101.htm; Internet accessed 2 February 2002.

240   "Air," *Agency for Toxic Substances & Disease Registry(ATSDR), Department of Health and Human Services (HHS)*, Washington, DC (n.d.) http://www.atsdr.cdc.gov/ general/theair.html; Internet accessed 26 July 2008.

241   "USA: US nuclear cost pips coal," Uranium Information Centre (UIC) Newsletter # 1, *Uranium Information Centre* (January-February 2000), at http://www.uic.com.au/ news101.htm; Internet accessed 2 February 2002.

242   Hans Holger Rogner and Arshad Khan, "Comparing Energy Options," *International Atomic Energy Agency* (n.d.): 1-5. http://www. iaea.or.at/worldatom/Periodicals/   Bulletin/Bull401/article1.html; Internet accessed 21 March 2002.

243   "Economics: energy and environmental analysis," *Los Alamos National Laboratory.* 27 April 2001, at http://www.lanl.gov/orgs/d/d4/ econo/economics.html; Internet accessed 7 April 2002.

244 "Consequence Based Analysis for Infrastructure Surety," Decision Applications Division, *Sandia National Laboratory*, at http://www.sandia.gov/Surety/Facts/ Consequence.htm; Internet accessed 7 April 2002..

245 "Nuclear Power," *International Energy Agency (IEA):Energy Technology Essentials,* Paris, March 2007.Ihttp://www.iea.org/textbase/techno/essentials4.pdf; Internet accessed 26 July 2008.

246 "GT-MHR Commercialization Study," General Atomics Project No. 30103, Final Technical Report for the Period June18, 2001 through June 30, 2002 (for U.S. Department of Energy; July 2002), at http://74.125.45.104/search?q=cache:dmq Wazb5UDQJ:www.osti.gov/bridge/servlets/purl/797074-xGJjKA/native/797074.pdf+gt mhr +acquisition+cost+per+kwe&hl=en&ct=clnk&cd=1&gl=us; Internet accessed 27 July 2008.

247 "The Economics of Nuclear Power," *World Nuclear Association,* London, July 2008, at http://www.world-nuclear.org/about/contact.html; Internet accessed 25 July 2008.

248 Walter A. Goetz, *The Contracting Process,* 6th ed. (Washington, DC: by the author, 1996), 40-43. Professor Goetz published this book to support a course in Contract Law offered by the Engineering Management and Systems Engineering Department, School of Engineering and Applied Science, George Washington University.

249 "The Economics of Nuclear Power," *World Nuclear Association,* London, July 2008, at http://www.world-nuclear.org/info/inf02.html; Internet accessed 30 July 2008.

250 Ibid.

251 "The Economics of Nuclear Power," 1-4

252 Ibid.

253 "Revival of Nuclear Power," *BHARAT RAKSHAK MONITOR,* Vol. 6(6), May-July 2004, at http://www.bharat-rakshak.com/MONITOR/ISSUE6-6/singh.html; Internet accessed 30 July 2008.

254   "The Economics of Nuclear Power," Nuclear Issues Briefing Paper # 8, *Uranium Information Centre* (January-February 2000), 2. http://www.uic.com.au/nip08.htm; Internet accessed 2 February 2002.

255   John Scire, "The answer to the waste storage problem, which is holding up nuclear power development, is to recycle the waste" (n.d.), at http://www.truthaboutenergy.com/ nucler%20fuel%20waste.htm; Internet accessed 30 July 2008.

256   "What the heck is a LUEC?" *Positive Energy,* 25 August 2004, at http://positiveenergy.blogspot.com/2004/08/what-heck-is-luec.html; Internet accessed 30 July 2008.

257   Moore and Guindon, 6.

258   M. R. Shay and others, "Life-Cycle Cost Analysis for Developing and Deploying a Representative Accelerator Transmutation of Waste System," *Pacific Northwest National Laboratory* (January 2000): 1-16.

259   "The Economics of Nuclear Power," 1-6.

260   "Cost and Benefits of Nuclear Power," *Canadian Nuclear FAQ* (n.d.), at http://www-formal.stanford.edu/ jmc/progress/notes.html#cohen; Internet accessed 8 April 2002.

261   Richard Rhodes and Denis Beller, "The Need for Nuclear Power, " *Foreign Affairs* (January/February 2000), 2.   http://www.nci.org/conf/rhodes/; Internet accessed 23 September 2002. R. Wilson and John Spengler, Eds. (1996). "Particles in Our Air: Concentrations and Health Effects." (Cambridge, Mass.: Harvard University Press), 212, at http://www.nci.org/conf/rhodes/; Internet accessed 26 September 2000. Paul Joskow and Edward Kahn, "A Quantitative Analysis of Pricing Behavior In California's Wholesale Electricity Market During Summer 2000," (a paper dated 15 January 2001), Massachusetts Institute of Technology (Cambridge, Mass.), 1-7, at http://econ-www.mit.edu/faculty/pjoskow/files/JK_PaperREVISED.pdf; Internet accessed 24 September 2002. Paul L. Joskow, "California's Electricity Crisis," (a paper dated 28 November 2001), Massachusetts Institute of Technology (Cambridge, Mass.), 1, at http://econwww.mit.edu/

faculty/pjoskow/files/CALIFORNIA_11-28-01.pdf; Internet accessed 25 September 2002.

262   "Acid Rain," Environmental Issues, *U.S. Environmental Protection Agency* (Washington, DC: 2001), 1-4.   http://epa.gov/airmarkets/acidrain; Internet accessed 21 March 2002.

263   Andrew Revkin, "New York to Sue Coal-Using Power Plants," *New York Times,* 15 September 1999, 1-3.   http://www.tngreen.com/air/News/nysuescoal.html; Internet accessed 21 March 2002.

264   Murray J. Stewart, "The Nuclear Option and Climate Change. A Necessary Part of Canada's Kyoto Implementation Strategy" (a paper presented at the 19th Annual Conference of the Canadian Nuclear Society, Toronto, Ontario: 19 October 1998), at http://www.cns-snc.ca/english/Speeches-Releases/presentations/Oct19-98.pdf;   Internet accessed 21 March 2002.

265   "Oil billionaire Pickens puts his money on wind power," *CNN.com/technology,* 8 July 2008.   http://www.cnn.com/2008/TECH/science/07/08/ pickens.plan/; Internet accessed 26 September 2008.

266   Eric Pianin, "Deaths Raise Alarm on Power Plants," *Washington Post.* 30 September 2001, A33.

267   Ibid.

268   Randi Fabi, "US takes aim at aging power plants for pollution," *Reuters Limited,* 5 November 1999.   http://www.climateark.org/articles/1999/ustakeai.htm; Internet accessed 25 September 2002.

269   Ibid.

270   Pianin, A33.

271   "New Commercial Reactor Designs," *Energy Information Administration: Official Energy Statistics from the U.S. Government,* November 2006, at http://www.eia.doe.gov/ cneaf/nuclear/page/analysis/nucenviss2.html; Internet accessed 30 July 2008.

272 "Nuclear Power," *Nuclear Electricity*, 6th ed., August 2000, *Uranium Information Centre* (updated: November 2001), Ch. 3, at http://www.uic.com.au//ne3.htm; Internet accessed 2 February 2002.

273 Alan Wong, 6. Wong points out that the economics of the GT-MHR are "very attractive" because of its high plant efficiency (16% higher than LWRs), modular construction, and simplicity of design.

274 Dave Nicholls, "The Pebble Bed Modular Reactor," *Nuclear News*, September 2001, 35.

275 Howard Eisner, *Essentials of Project and Systems Engineering Management* (New York: John Wiley & Sons, Inc., 1997), 210-211.

276 Bartlett's, 65.15.

277 Jim Vande Hei, "A High-Energy Report," *Wall Street Journal*, 18 May 2001, A6.

278 David Krieger, *Bush-Putin Nuclear Arms Cuts Are Not Enough*, Nuclear Age Peace Foundation, November 2001, at http://www.wagingpeace.org/articles/2001/11/00_krieger_bush-putin.htm; Internet accessed 30 July 2008.

279 "Disarmament and Arms Control Treaties," *Reaching Critical Will*, Women's International League for Peace and Freedom (n.d.) at http://www.reachingcriticalwill.org/legal/treaties.html; Internet accessed 30 July 2008.

280 Siegfried S. Hecker, "Thoughts about an Integrated Strategy for Nuclear Cooperation with Russia," *The Nonproliferation Review* (Summer 2001) Vol. 8, No. 2, at http://cns.miis.edu/pubs/npr/vol08/82/heck82.htm; Internet accessed 29 September 2008.

281 Charles Krauthammer, "War and the Polar Bear," *The Washington Post*, 9 November 2001. Also available at *Jewish World Review*, 9 November 2001, at http://www.jewish worldreview.com/cols/krauthammer110901.asp; Internet accessed 11 August 2008.

282 Will Lester, *Poll: US Nuke Power Anxiety Easing April 25, 2001*, International Communications Research (n.d.) at http://www.

icrsurvey.com/Study.aspx?f=AP_ Nukepoll.html; Internet accessed 30 July 2008.

283 "World Resources 1996-97, A Guide to the Global Environment," *World Resources Institute* (Washington, DC, 1997), Ch. 12, at http://www.wri.org/wri/wr-96-97/em_txt4.html; Internet accessed 10 October 2002.

284 Ibid., Ch. 8

285 "Bush Approves Yucca Nuclear Waste Site," *Online News Hour,* 15 February, 2002. http://www.pbs.org/newshour/updates/february02/nuclear_2-15.html; Internet accessed, 31 July 2008.

286 Steve Hargreaves, *World energy use seen surging,* CNN Money Special Report, 25 June 2008, at http://money.cnn.com/2008/06/25/news/economy/eia_outlook/index.htm? postversion=2008062512; Internet accessed 31 July 2008.

287 "Nuclear Power in Italy," *World Nuclear Association,* London, August 2008, at http://www.world-nuclear.org/info/inf101.html; Internet accessed 17 August 2008.

288 Janice Forsythe, D.Sc., "Nuclear Waste: Asset or Liability? A Pragmatic View in the 21st Century (Doctoral dissertation defended 22 April, 2003) Ch 2, 57-58, at http://www.google.com/search?hl=en &q=Janice+I.+Forsythe+dissertation+nuclear+waste; Internet accessed 17 August 2008.

289 Andrea Cappannari, *California's energy debacle continues: From power shortage to power surplus,* Large & Associates, London, 4 January 2002, at http://www.wsws.org/ articles/2002/jan2002/cali-j04_prn.shtml; Internet accessed 17 August 2008.

290 Senator Frank Murkowski, presenter at the "Energy for Tomorrow's World," Nuclear Energy Assembly, Washington, DC, 22 May 2001, at http://www.cogema-inc.com/news/releases/05_22_01.shtm; Internet accessed 28 May 2002.

291 "Advanced Reactors," Nuclear Issues Briefing Paper #16, *Uranium Information Center* (Melbourne, Australia, November 2002), 1. http://www.uic.com.au/nip16.htm; Internet accessed 21 January 2003.

292 Anne Lauvergeon, "Energy for Tomorrow's World," Nuclear Energy Assembly, Washington, DC, 22 May 2001, at http://www.cogemainc.com/news/releases/ 05_22_01.shtm; Internet accessed 28 May 2002.

Note: For more information on nuclear technology, issues, and news, the following Internet sources are particularly rich:

World Nuclear Association, London

International Atomic Energy Agency

World Nuclear News

Virtual Nuclear Tourist

CASEnergy Coalition (USA)

Nuclear Energy Institute, USA

World Energy Council.

Australian Nuclear Science & Technology Organisation (ANSTO)

Nuclear Fuel Cycle Information Service (directory, IAEA)

OECD Nuclear Energy Agency

US Energy Information Administration section on uranium & nuclear industry stats

Canadian Nuclear FAQ

World Nuclear Transport Institute

British Energy (major UK utility)

Sellafield Ltd (UK nuclear fuel cycle management)

WasteLink guide to resources on radioactive waste

UN Scientific Committee on the Effects of Atomic Radiation (UNSCEAR)

New Zealand Atomic Energy Advocacy Council

SELECTED GLOSSARY
Abbreviations, Acronyms, and Definitions

ABM	Antiballistic Missile
ABWR	Advanced Boiling Water Reactor
ACDA	Arms Control and Disarmament Agency
AEA	(U.S.) Atomic Energy Act
AEC	(U.S.) Atomic Energy Commission
APS	Arizona Public Services
BTU	British Thermal Unit
BWR	Boiling Water Reactor
CA	Construction Authorization
CANDU	Canadian Deuterium Uranium (Reactor)
CEA	Commissariat à l'Energie Atomique (the French Atomic Energy Commission)
CVN	Aircraft Carrier (CV) Nuclear (N)
D&D	Decommission and Decontamination
DCS	Duke, Cogema, Stone & Webster
DNFSB	Defense Nuclear Facilities Safety Board
DoD	Department of Defense
DOE	Department of Energy
EDF	Electricite de France
EIA	Energy Information Administration
EPA	Environmental Protection Agency
EUAC	Equivalent Uniform Annual Cost
FBR	Fast Breeder Reactor
FEIS	Final Environmental Impact Statement

FERC	Federal Energy Regulatory Commission
FNR	Fast Neutron Reactor
GT-MHR	Gas Turbine-Modular Helium Reactor
GW	Gigawatt
GWd/MT	Gigawatt Days per Metric Ton
GWh	Gigawatt Hour
HEU	High-Enriched Uranium
HLW	High-Level Waste
HWR	Heavy Water Reactor
IAEA	International Atomic Energy Agency
IEER	Institute for Energy and Environmental Research
INEEL	Idaho National Engineering and Environmental Laboratory
INES	International Nuclear Event Scale
ISO	International Organization for Standardization
JCS	Joint Chiefs of Staff
LA	License Application
LANL	Los Alamos National Laboratory
LEU	Low-Enriched Uranium
LFTR	Liquid Fluoride Reactor
LLNL	Lawrence Livermore National Laboratory
LLW	Low-Level Waste
LMR	Liquid Metal Reactor
LWR	Light Water Reactor (BWR & PWR are collectively known as LWRs)
MFFF	MOX Fuel Fabricating Facility
MOU	Memorandum of Understanding
MOX	Mixed-Oxides Fuels
mrem	Millirem, a measurement of radiation exposure
mSv	Millisievert, a measurement of radiation exposure

MTHM	Metric Tons of Heavy Metal
MWe	Megawatts (Electric) output from a generator
MWh	Megawats per Hour
MWt	Megawatts (Thermal) output from a reactor
MTU	Metric Tons Uranium
NAS	National Academy of Sciences
NEA	Nuclear Energy Agency
NEI	National Energy Institute
NNPP	Naval Nuclear Propulsion Program
NNPS	Notional Nuclear Power System
NNSA	National Nuclear Security Agency
NPP	Nuclear Power Plant
NRC	(U.S.) Nuclear Regulatory Commission
NRP	Naval Reactor Program
O&M	Operation and Maintenance
ONRL	Oak Ridge National Laboratory
OPRI	Office for Protection Against Radiation Ionizing
PBMR	Pebble Bed Modular Reactor
PNNL	Pacific Northwest National Laboratory
PRA	Probabilistic Risk Assessment
PWR	Pressurized Water Reactor
rem	Roentgen Equivalent Man
R&D	Research and Development
RTG	Radioisotope Thermoelectric Generators
SNL	Sandia National Laboratories
SRS	Savannah River Site
SSN	Submarine (SS) Nuclear (N)
STR	Submarine Thermal Reactor
T&D	Transmission and Distribution
TMI	Three Mile Island

TVA	Tennessee Valley Authority
UCS	Union of Concerned Scientists
UDI	Utility Data Institute
UIC	Uranium Information Center
USEC	United States Enrichment Corporation
WNA	World Nuclear Association
WPu	Weapons-Grade Plutonium
WU	Weapons-Grade Uranium

DEFINITIONS

Acid Rain: *Wet* acid falls out of the atmosphere in the form of rain, fog, and snow; the process is called deposition. *Dry* acid consists of acidic gases and particles. Acid rain affects living things, materials, and human health.

Alpha Particle: A positively charged particle from the nucleus of an atom, emitted during radioactive decay. Alpha particles are helium nuclei, with 2 protons and 2 neutrons.

Americium: A white metallic synthetic element of the actinide series. It is the longest lived isotope.

Ampere: Steady current produced by one volt applied across a resistance of one ohm.

Atom: A particle of matter that cannot be broken up by chemical means. Atoms have a nucleus consisting of positively charged protons and uncharged neutrons of the same mass. The positive charges on the protons are balanced by a number of negatively charged electrons in motion around the nucleus.

Background Radiation: The naturally-occurring ionizing radiation in our environment, to which every person is exposed, arising from the earth's crust and from cosmic radiation.

In Virginia, the exposure level is approximately 100 millirem per year.

Beta Particle: A particle emitted from an atom during radioactive decay. Beta particles may be either electrons (with negative charge) or positrons.

Biological Shield: A mass of absorbing material (e.g., thick concrete walls) placed around a reactor or radioactive material to reduce the radiation (especially neutrons and gamma rays respectively) to a level safe for humans.

Boiling Water Reactor: A light-water reactor (LWR) in which water, used both as a coolant and moderator, is allowed to boil in the core. The resulting steam can be used directly to drive a turbine.

Breed: To form fissile nuclei, usually as a result of neutron capture, possibly followed by radioactive decay.

Breeder Reactor: A reactor that produces more fissile material than it consumes.

British Thermal Unit: The heat equal to 1/180 of the heat required to raise the temperature of one pound of water from 32o to 212 oF at a constant pressure of one atmosphere.

Burnup: A measure of the neutron irradiation of the nuclear fuel. It is normally quoted in megawatt–days per metric ton of uranium metal or its equivalent (MWd/MTU), or gigawatt–days/MTU (GWd/MT).

CANDU: The Canadian deuterium uranium reactor, moderated and cooled by heavy water.

Chain Reaction: A reaction that stimulates its own repetition, in particular where the neutrons originating from nuclear fission cause an ongoing series of fission reactions.

Cladding: The metal tubes containing oxide fuel pellets in a reactor core.

Commercial Operation: A generating unit is said to be in commercial operation when control of the loading of the unit is turned over to the system dispatcher.

Control Rods: Devices to absorb neutrons so that the chain reaction in a reactor core may be slowed or stopped by inserting them further, or accelerated by withdrawing them.

Consumption (Fuel): The amount of fuel used for gross generation, providing standby service and start-up and/or flame stabilization. (See Fuel)

Coolant: The liquid or gas used to transfer heat from the reactor core to the steam generators or directly to the turbines.

Core: The central part of a nuclear reactor containing the fuel elements and any moderator.

Critical Mass: The smallest mass of fissile material that will support a self-sustaining chain reaction under specified conditions.

Criticality: Condition of being able to sustain a nuclear chain reaction.

Current: A flow of electrons in an electrical conductor. The strength or rate of movement of the electrons is measured in amperes. (See Ampere, Ohm, Volt.)

Decay: Disintegration of atomic nuclei resulting in the emission of alpha or beta particles (usually with gamma radiation). Also the exponential decrease in radioactivity of a material as nuclear disintegrations take place and more stable nuclei are formed.

Decommissioning: Removal of a facility (e.g., reactor) from service, also the subsequent actions of safe storage, dismantling, and making the site available for unrestricted use.

Demand: The rate at which electric energy is delivered to or by a system, part of a system, or piece of equipment, at any given instant or averaged over any designated period of time.

Depleted Uranium: Uranium having less than the natural 0.7% U-235.

Design Electrical Rating (Capacity), Net: The nominal net electrical output of a nuclear unit, as specified by the utility for plant design.

Deuterium: "Heavy Hydrogen," a stable isotope having one proton and one neutron in the nucleus.

Dirty Bomb: Made by attaching radioactive nuclear material to the exterior of a conventional bomb or other explosive device. There is no nuclear explosion; the nuclear material is simply spread by the explosion of the conventional weapon.

Dose: The energy absorbed by tissue from ionizing radiation.

Electric Plant: A station containing prime movers, electric generators, and auxiliary equipment for converting mechanical, chemical, and/or fission energy into electric energy.

Electric Power Industry: The public, private, and cooperative electric utility systems of the United States taken as a whole. This includes all electric systems serving the public: regulated investor-owned electric utility companies; Federal power projects; State, municipal, and other government-owned systems, including electric public utility districts; electric cooperatives, including Generation and Transmission entities; jointly owned electric utility facilities, and electric utility facilities owned by a lessor and leased to an electric utility. Excluded from this list are the special purpose electric facilities or systems that do not offer service to the public.

Electric Power System: An individual electric power entity – a company, an electric cooperative, a public electric supply corporation like the Tennessee Valley Authority, a similar

Federal department or agency like the Bonneville Power Administration, the Bureau of Reclamation or the Corps of Engineers, a municipally-owned electric department offering service to the public, or an electric public utility district ("PUD"); also, a jointly-owned electric supply project such as the Keystone.

Electric Utility: A corporation, person, agency, authority, or other legal entity or instrumentality that owns and operates facilities within the United States, its territories, or Puerto Rico for the generation, transmission, distribution, or sale of electric energy, primarily for use by the public. An entity that solely operates qualifying facilities under the Public Regulatory Policies Act of 12978 is not considered an electric utility.

Element: A chemical substance that cannot be divided into simple substances by chemical means; atomic species with the same number of protons.

Energy: The capacity for doing work as measured by the capability of doing work (potential energy) or the conversion of this capability to motion. Energy has several forms, some of which are easily convertible and can be changed to another form useful for work. Most of the world's convertible energy comes from fossil fuels that are burned to produce heat that is then used as a transfer medium to mechanical or other means in order to accomplish tasks. Electrical energy is usually measured in kilowatt hours, while heat energy is usually measured in British thermal units (BTUs).

Energy Source: The primary source that provides the power that is converted to electricity through chemical, mechanical, or other means. Energy sources include coal, petroleum and petroleum products, gas, water, uranium, wind, sunlight, geothermal, and other sources.

Enriched Uranium: Uranium in which the proportion of U-235 (to U-238) has been increased to more than the natural 0.7%.

Enrichment: Physical process of increasing the proportion of U-235 to U-238.

External Cost: A cost, e.g., of waste disposal or environmental impact, that is omitted when calculating the cost, price, or economic benefit of a product or service.

Fast Breeder Reactor (FBR): A fast neutron reactor configured to produce more fissile material than it consumes, using fertile material such as depleted uranium in a blanket around the core.

Fissile: Fissionable, especially by neutrons.

Fission: A nuclear reaction in which an atomic nucleus, especially a heavy nucleus, splits into fragments, usually two fragments of comparable mass, with the release of energy.

Fossil Fuel: A fuel based on carbon presumed to be originally from living matter, e.g., coal, crude oil, and natural gas, which is burned (with oxygen) to yield energy.

Fuel: Any substance that can be burned to produce heat; also materials that can be fissioned in a chain reaction to produce heat.

Fuel Assembly: Structured collection of fuel rods or elements, the unit of fuel in a reactor.

Gamma Rays: High energy electro-magnetic radiation, virtually identical to x-rays.

Generating Unit: An electric generator together with its prime mover.

Generation: The process of producing electric energy by transforming other forms of energy; also, the amount of electric energy produced, expressed in kilowatt hours.

Generator: A machine that converts *mechanical* (emphasis added) energy into electrical energy. (Note that a reactor is not properly referred to as a generator.)

Gigawatt (GW): One billion watts or one million kilowatts.

Gigawatt hour (GWh): One billion watt hours.

Greenhouse Effect: The phenomenon whereby the earth's atmosphere traps solar radiation, caused by the presence of gases such as carbon dioxide that allow incoming sunlight to pass through but absorb heat radiated back from the earth's surface.

Greenhouse Gases: Radiative gases in the earth's atmosphere that absorb long-wave heat radiation from the earth's surface and re-radiate it, thereby warming the earth. Carbon dioxide and water vapor are the main ones.

Grid: The layout of an electrical distribution system.

Half-Life: The period required for half of the atoms of a particular radioactive isotope to decay and become an isotope of another element.

Heavy-Water: Water containing an elevated concentration of molecules with deuterium ("heavy hydrogen") atoms.

Heavy-Water Reactor (HWR): A reactor that uses heavy water as its moderator.

High-Level Wastes: Extremely radioactive fission products and transuranic elements (usually other than plutonium) in spent nuclear fuel. Requires both shielding and cooling, and very careful handling, storage, and disposal. It can be recovered by reprocessing spent fuel; some countries regard spent fuel itself as HLW.

Highly-Enriched Uranium (HEU): Uranium enriched to at least 20% U-235. (That in weapons is about 90% U-235.)

In Situ Leaching (ISL): Recovery by chemical leaching of minerals from porous orebodies without physical excavation. Also known as solution mining.

Ion: An atom that is electrically charged because of the loss or gain of electrons.

Intermediate-Level Waste (ILW): Sufficiently radioactive to require shielding, some is categorized as long lived ILW, and this may be disposed of with HLW.

Internal Cost: A cost, e.g., of waste disposal or environmental impact that is taken into account when calculating the cost, price, or economic benefit of a product or service.

Isotope: An atomic form of an element having a particular number of neutrons. Different isotopes of an element have the same number of protons but different numbers of neutrons, hence different atomic mass, e.g., U-235, U238.

Joule: A unit of electrical energy equal to the work done when a current of 1 ampere is passed through a resistance of 1 ohm for 1 second.

Kilowatt (kW): One thousand watts.

Kilowatt hour: One thousand watt hours.

Life Extension: Investments made to maintain the operating status of an electric generating plant into acceptable levels of availability and efficiency beyond its originally anticipated retirement date.

Light-Water Reactor (LWR): A common nuclear reactor cooled and usually moderated by ordinary water.

Load (Electric): The amount of electric power delivered or required at any specific point or points on a system. The requirement originates at the energy-consuming equipment of the customers.

Load Management Technique: Utility demand management practices directed at reducing the maximum kilowatt demand on an electric system, and/or modifying the coincident peak demand of one or more classes of service to better meet the utility system capability for a given hour, day, week, season, or year.

Low-Enriched Uranium: Uranium enriched to less than 20% U-235. (That in power reactors is usually 3.5 – 4.5% U-235.)

Low-Level Waste (LLW): Mildly radioactive material that does not require shielding in handling or storage, and is usually disposed of by incineration and shallow burial.

Megawatt (MW): A unit of power, one million watts. MWe refers to electric output from a generator, MWt to thermal output from a reactor or heat source (e.g., the gross heat output of a reactor itself, typically three times the MWe figure).

Megawatt hour (MWh): One million watt hours.

Metal Fuels: Natural uranium metal as used in a gas-cooled reactor.

Milling: Process by which minerals are extracted from ore, usually at the mine site.

Millirem: A unit used to measure radiation dosage. It is 1/1000 of a rem, which is a unit used to measure radiation dosage. It also relates to the potential effect of radiation on human cells.

Mixed-Oxide Fuel (MOX): Reactor fuel that consists of both uranium and plutonium oxides, usually about 5% Pu, which is the main fissile component.

Moderator: A material such as light or heavy water or graphite used in a reactor to slowdown fast neutrons by collision with lighter nuclei so as to expedite further fission.

Natural Uranium: Uranium with an isotopic composition as found in nature, containing 99.3% U-238, 0.7% U-235, and a trace of U-234. Can be used as fuel in heavy water-moderated reactors.

Net Generation: Gross generation less plant use, measured at the high-voltage terminals of the station's step-up transformer. The energy required for pumping at pumped-storage plants is regarded as plant use and must be deducted from the gross generation.

Neutron: An uncharged elementary particle found in the nucleus of every atom except hydrogen. Solitary mobile neutrons traveling at various speeds originate from fission reactions. Slow (thermal) neutrons can readily cause fission in nuclei of "fissile" isotopes, e.g., U-235, Pu-239, U-233; and fast neutrons can cause fission in nuclei of "fertile" isotopes such as U-238 and Pu-239. Sometimes atomic nuclei simply capture neutrons.

Nuclear Fuel: Fissionable materials that have been enriched to such a composition that when placed in a nuclear reactor will support a self-sustaining fission chain reaction, producing heat in a controlled manner for process use.

Nuclear Power Plant: A plant in which the prime mover is a turbine. The steam used to drive the turbine is produced by a heat transfer from the reactor vessel during the period when the nuclear fuel is undergoing fission.

Nuclear Reactor: A device in which a nuclear fission chain reaction occurs under controlled conditions so that the heat yield can be harnessed or the neutron beams utilized. All commercial reactors are thermal reactors, using a moderator to slowdown the neutrons.

Nuclear Regulatory Commission: The federal agency responsible for the regulation and inspection of nuclear power plants to assure safety.

Ohm: Unit of electric resistance equal to the resistance of a circuit in which a potential difference of one volt produces a current of one ampere.

Operable: A unit is operable when it is available to provide power to the grid. For a nuclear unit, this is when it receives its full power amendment to its operating license from the Nuclear Regulatory Commission.

Other Generation: Electricity originating from these sources: biomass, fuel cells, geothermal heat, solar power, waste, wind, and wood.

Outage: A period of interruption of electric current.

Oxide Fuels: Enriched or natural uranium in the form of the oxide UO_2, used in many types of reactors.

Peak Load: The maximum load during a specified period of time.

Peaking Capacity: Capacity of generating equipment normally operated during the hours of highest daily, weekly, or seasonal loads. Some generating equipment may be operated at certain times as peaking capacity and at other times to serve loads on a 'round-the-clock' basis.

Plant: A station at which are located prime movers, electric generators, and auxiliary equipment for converting mechanical, chemical, and/or nuclear energy into electric energy. A station may contain more than one type of prime mover.

Plutonium: A transuranic element, formed in a nuclear reactor by neutron capture. It has several isotopes, some of which are fissile and some of which undergo spontaneous fission, releasing neutrons. Weapons-grade plutonium is produced in special reactors to give >90% Pu-239; reactor-grade plutonium contains about 30% non-fissile isotopes. About one third of the energy in a light-water reactor comes from the fission of Pu-239, and this is the main isotope of value in reprocessing spent fuel.

Power: A source or means of supplying energy, especially electricity.

Power (Electrical): An electric measurement unit of power called a volt-ampere is equal to the product of one volt and one ampere. This is equivalent to 1 Watt for a direct current system and a unit of apparent power is separated into real and reactive power. Real power is the work-producing part of apparent power that measures the rate of supply of energy and is denoted as kilowatts (KW). Reactive power is the portion of apparent power that does no work and is referred to as kilovars; this type of power must be supplied to most types of magnetic equipment, such as motors, and is supplied by generator or by electrostatic equipment. Volt-amperes are usually divided by 1,000 and called kilovolt-amperes (kVA). Energy is denoted by the product of real power and the length of time utilized; this product is expressed as kilowatt hours.

Pressurized-Water Reactor (PWR): The most common type of light-water reactor (LWR), it uses water at very high pressure in a primary circuit, and steam is formed in a secondary circuit.

Prime Mover: The engine, turbine, water wheel, or similar machine that drives a generator.

Privately Owned Electric Utility: A class of ownership found in the electric power industry where the utility is regulated and authorized to achieve an allowed rate of return.

Production (Electric): Act or process of producing elect4ric energy from other forms of energy; also, the amount of electric energy expressed in watt hours (Wh).

Publicly Owned Electric Utility: A class of ownership found in the electric power industry. This group includes those utilities operated by municipalities, and State and Federal power agencies.

Rad: A unit of energy absorbed from ionizing radiation, equal to 100 ergs per gram or 0.01 joule per kilogram of irradiated material.

Radiation: Energy given off in the form of waves or particles. The term "radiation" is broad and includes ordinary sunlight and radio waves, but more often it is used to mean "ionizing" radiation. Ionizing radiation can produce charged particles in materials that it strikes, including living matter. The most common types of ionizing radiation are alpha, beta, and gamma.

Alpha is the least penetrating type of radiation; it can be stopped with a sheet of paper.

Beta radiation is emitted from the nucleus of an atom during fission; it consists of electrons that can be stopped by a thin cardboard.

Gamma radiation is electro-magnetic waves emitted from the nucleus of an atom and is essentially the same as x-rays. It can be stopped with heavy shielding, such as concrete or lead.

Radioactivity: The spontaneous decay of an unstable atomic nucleus, giving rise to the emission of radiation.

Radionuclide: A radioactive isotope of an element.

Radiotoxicity: The adverse health effect of a radionuclide due to its radioactivity.

Reactor Pressure Vessel: The main steel vessel containing the reactor fuel, moderator, and coolant.

Rem: The amount of ionizing radiation required to produce the same biological effect as one rad of high-penetration x-rays.

Repository: A permanent disposal place for radioactive wastes.

Reprocessing: Chemical treatment of spent fuel (from reactors) to separate uranium and plutonium from the small quantity of fission product waste products and transuranic elements, leaving a much reduced quantity of high-level waste.

Sintering: To form a coherent mass by heating without melting.

Spent Fuel: Fuel assemblies removed from a reactor after several years use.

Standby Facility: A facility that supports a utility system and is generally running under no-load. It is available to replace or supplement a facility normally in service.

Station (Electric): A plant containing prime movers, electric generators, and auxiliary equipment for converting mechanical, chemical, and/or nuclear energy into electric energy.

System (Electric): Physically connected generation, transmission, and distribution facilities operated as an integrated unit under one central management, or operating supervision.

Thermal: A term used to identify a type of electric generating station, capacity, capability, or output in which the source of energy for the prime mover is heat.

Thermal Reactor: A nuclear reactor in which the fission chain reaction is sustained primarily by slow neutrons, hence requiring a moderator (as distinct from a fast neutron reactor).

Thermalized: Lowered in energy. A fast neutron loses energy as it collides with light elements during the fission process.

Transuranic Element: A very heavy element formed artificially by neutron capture and subsequent beta decay. Has a higher atomic number than uranium (92). All are radioactive; neptunium, plutonium, and americium are the best-known.

Turbine: A rotary engine actuated by the reaction or impulse or both of a current of fluid (as water, steam, or air) subject to pressure; usually made with a series of curved vanes on a central rotating spindle.

Uranium (U): A mildly radioactive element with two isotopes that are fissile (U-235 and U-233) and two that are fertile (U-238 and U-234). Uranium is the basic fuel of nuclear energy.

Vitrification: The incorporation of high-level wastes into borosilicate glass, to make up about 14% of it by mass. It is designed to immobilize radionuclides in an insoluble matrix ready for disposal.

Volt: The difference of potential between two points in a conducting wire carrying a constant current of one ampere when the power dissipated between these two points is equal to one watt and equivalent to the potential difference across a resistance of one ohm when one ampere is flowing through it.

Waste: See: High-Level Waste (HLW) and Low-Level Waste.

Watt: Unit of (electrical) power equal to the rate of work represented by a current of 1 ampere under a pressure of 1 volt.

Watt-hour: Energy equivalent to the power of one watt operating for one hour.

EXPLANATION OF RADIATION

Average annual effective dose equivalent to persons in the United States from various radiation sources:

Man-made (measured in millirem-per-year)
Medical:

Diagnostic X-rays	39.00
Nuclear medicine	14.00
Consumer products	5 - 13.00
Occupational	0.90
Miscellaneous:	
Environmental	0.06
Nuclear fuel cycle	0.05

Natural background (measured in millirem-per-year)

Radon	200.00
Cosmic rays	27.00
Cosmogenic radiation	1.00
Terrestrial radiation	28.00
Internal radiation	39.00

Radiation Protection

You cannot see or smell radiation; it can be detected accurately with the aid of instruments designed for that purpose. Trained technicians using these instruments monitor radiation in and around nuclear plants. The public is protected because, should a nuclear incident occur, this monitoring will be increased to obtain accurate information for all areas affected.

Radiation

> There is nothing new regarding radiation. It has always been a part of our natural environment. We are constantly exposed to radiation from the sun and outer space.

Radioactive materials exist in the earth around us, in the buildings we live and work in, and in the food and water we consume. There are radioactive gases in the air we breathe, and our bodies themselves are radioactive. Nuclear power is a small contributor to the average radiation exposure. Persons living in Denver, Colorado, receive double the amount of natural radiation that we receive in Virginia because of Denver's higher altitude. We are also exposed to sources of man-made radiation. For more than half a century, doctors and scientists have used X-rays and other forms of penetrating radiation. Medical diagnosis and treatment are the main sources of exposure to man-made radiation, but the benefits far outweigh the problems.

Within a decade after the X-ray came into use, it was apparent that it could be either beneficial or harmful, depending on its use and control and the protective measures that would be necessary. This applies to other uses of nuclear energy as well, including the power industry.

The fission process that takes place in a nuclear power plant is a source of man-made radiation, although in normal operations, the amount reaching the environment is rather insignificant. The average person receives approximately 350 millirems per year from natural and man-made sources, and a person living within 10 miles of a nuclear generating plant receives less than one millirem each year from that plant, because it is designed and built to prevent radioactivity from reaching the environment, both during normal operation and in the event of an accident. These intensive efforts by the industry have worked in the more than 25 years of nuclear power production in the United States. Not a single death or serious injury from radiation has ever been recorded involving a member of the public. The likelihood of such an occurrence in the future is *de minimis*.

BIBLIOGRAPHY

Books

Aguilar, Francis Joseph. *General Managers in Action—Policies and Strategies,* 2nd ed. New York: Oxford University Press, 1992.

Cameron, Gavin. "Nuclear Terrorism: A Threat Assessment for the 21st Century," New York: St. Martins Press, 1999.

Cascio, Wayne F. *Managing Human Resources—Productivity, Quality of Work Life, Profits,* 2nd ed. New York: McGraw-Hill, 1989.

Chandler, Alfred D., Jr. *Scale and Scope. The Dynamics of Industrial Capitalism.* Cambridge: Harvard University Press.

Davenport, Thomas H. and Laurence Prusak. *Working Knowledge— How Organizations Manage What They Know.* Boston: Harvard Business School Press, 2000.

Eisner, Howard. *Essentials of Project and Systems Engineering Management,* New York: John Wiley & Sons, Inc., 1997.

Goodstein, David L. *Out Of Gas: The End of the Age of Oil,* New York: W. W. Norton & Company, 2005

Hart, Christopher W. L. and Christopher E. Bogan. *The Baldrige—What It Is, How Its Won, How To Use It To Improve Quality In Your Company.* Boston: McGraw-Hill,1992.

Hore-Lacy, Ian. *Nuclear Energy in the 21st Century.* London: World Nuclear University Press, 2006.

Johansson, Johny K. *Global Marketing,* 2nd ed. Boston: McGraw-Hill, 2000.

Kanter, Rosabeth Moss. e ·*Volve! Succeeding in the Digital Culture of Tomorrow.* Boston: Harvard Business School Press, 2001.

Keegan, Warren J. *Global Marketing,* 6th ed. Upper Saddle River, N. J.: Prentice Hall, 1999.

Kepner, Charles H. and Benjamin B. Tregoe. *The Rational Manager.* Akron: Kepner-Tregoe, Inc., 1976.

Kuratko, Donald F. and Richard M. Hodgetts. *Entrepreneurship. A Contemporary Approach,* 5th ed. Fort Worth: Harcourt, 2001.

Lavoy, Peter R.; Sagan, Scott Douglas; and Wirtz, James J. [eds.]. "Planning the unthinkable: how new powers will use nuclear, biological, and chemical weapons." Ithaca: Cornell University Press, (Cornell Studies in Security Affairs), 2000.

Martin, Daniel, "Three Mile Island -- Prologue or Epilogue?" Cambridge: Ballinger Publishing Company, 1980.

Morison, Samuel Eliot. *History of Naval Operations in World War II.* Vol. IV, *Coral Sea, Midway and Submarine Actions, May 1942-August 1942.* Boston: Little, Brown and Company, 1949.

Newstrom, John W. and Keith Davis. *Organizational Behavior. Human Behavior at Work,* 10th ed. Boston: McGraw-Hill, 1997.

Risley, George. *Modern Industrial Marketing. A Decision-Making Approach.* New York: McGraw-Hill, 1972.

Robbins, Stephen P. *Organizational Behavior,* 9th ed. Upper Saddle River, N. J.: Prentice Hall, 2001.

Senge, Peter M. *The Fifth Discipline. The Art & Practice of The Learning Organization.* New York: Doubleday, 1990.

Slater, Robert. *The GE Way Field Book. Jack Welch's Battle Plan for Corporate Revolution.* New York: McGraw- Hill, 2000.

Sullivan, William G., James A. Bontadelli, and Elin M. Wicks. "Engineering Economy," 11th ed., Upper Saddle River, NJ: Prentice Hall, 1997.

Tiwana, Amrit. *The Knowledge Management Toolkit.* Upper Saddle River, N. J.: Prentice Hall, 2000.

Thurow, Lester. *Head to Head—The Coming Economic Battle Among Japan, Europe, and America.* New York: William Morrow, 1992.

Journal Articles, Periodicals, and Newspapers

Baier, Donald. "The Harrisburg Hoax," *Fusion,* 2 (May) 1972.

Bradley, William A. [et al.]. "Keeping Reactors Safe from Sabotage," *Los Alamos Science,* (Summer-Fall) 1981.

Denton, Jeremiah. "International Terrorism: The Nuclear Dimension." *Terrorism,* 9:2, 1987.

Falkenrath, Richard A. "Problems of Preparedness: U.S. Readiness for a Domestic Terrorist Attack," *International Security,* 25:4 (Spring) 2001.

Hirsch, Daniel, [et al.]. "Protecting Reactors from Terrorists," *Bulletin of the Atomic Scientists,* 42:3 (March) 1986.

Jenkins, Brian M. "International Cooperation in Locating and Recovering Stolen Nuclear Materials," *Terrorism,* 6:4 1983.

"Will Terrorists Go Nuclear?" *Orbis,* 29:3 1985.

Jones, Arthur. "Nuclear Terrorism 'More Likely' Than War; Experts Say U.S. Likely Target," *National Catholic Reporter,* 7 (February) 1986.

Mintzberg, Henry 1981. "Organizational Design: Fashion or Fit?" *Harvard Business Review* (January-February)

"Jane's Fighting Ships" 1982-83, ed. Captain John Moore RN, Jane's Publishing Company Limited, London, 1982.

Post, Jerold M. "Prospects for Nuclear Terrorism: Psychological Motivations and Constraints," *Conflict Quarterly,*7 (Summer) 1987.

Ruthen, Russell. "Iraq's Nuclear Threat: Is the U.S. Underestimating Hussein's Nuclear Capability?" *Scientific American,*264 (February) 1991.

Sopko,John F. "The Changing Proliferation Threat," *Foreign Policy,* 105 (Winter) 1996-1997.

Vegar, Jose. "Terrorism's New Breed," *Bulletin of the Atomic Scientists,* 54:2 (March-April) 1998.

Useful Internet Sources

World Nuclear Association, London including Symposium Proceedings and portal.

International Atomic Energy Agency

World Nuclear News comprehensive news service

Nuclear Science and Technology and How It Influences Your Life - American Nuclear Society. See also their public information section.

Virtual Nuclear Tourist a very comprehensive information source, including directory of reactors.

CASEnergy Coalition (USA) - promoting clean and safe energy.

Nuclear Energy Institute, USA, major industry body.

Federation of Electric Power Companies of Japan, major industry body.

World Energy Council.

Australian Safeguards & Non-proliferation Office (ASNO), especially Current Topics and Background sections of Annual Reports

Australian Nuclear Science & Technology Organisation (ANSTO)

Nuclear Fuel Cycle Information Service (directory, IAEA)

OECD Nuclear Energy Agency

US Energy Information Administration section on uranium & nuclear industry stats

Canadian Nuclear FAQ

World Nuclear Transport Institute

British Energy (major UK utility)

Sellafield Ltd (UK nuclear fuel cycle management)

WasteLink guide to resources on radioactive waste

UN Scientific Committee on the Effects of Atomic Radiation (UNSCEAR)

Australian House of Reps Inquiry into Importance of Australia's Uranium Resources Australian Uranium Mining, Processing and Nuclear Energy Review report Dec 2006 (Switkowski Report).

New Zealand Atomic Energy Advocacy Council

ABOUT THE AUTHOR

Jan Forsythe completed all doctoral course work required for the degree Doctor of Philosophy in Public Administration at the University of Colorado and earned the Doctor of Science degree in Engineering Management at George Washington University (GWU). Her dissertation was entitled *Nuclear Waste: Asset or Liability? A Pragmatic View in the 21st Century* (Copyright 2003). France that went nuclear as the result of the energy crisis in 1973 and the U.S. Navy that embraced nuclear power under the leadership of then-Captain Admiral Hyman Rickover in 1948, served as models for her research. Prior to her work in the nuclear field, Dr. Forsythe taught school in Virginia, Delaware, the Philippines, and Colorado, where she was nominated as Colorado's Teacher of the Year (1976) in a national contest. While at GWU in the mid-1980s, where she was on the staff of the Institute for Technology and Strategic Research (ITSR) headed by Dr. Edward Teller, she changed her career field to nuclear energy. She has worked for EG&G, SAIC, Lockheed Martin, and Bechtel, companies that successively managed the Idaho National Laboratory (INL). Dr. Forsythe served in Washington D.C. representing INL as liaison to the Congress and the Department of Energy, and she spent several years "in the field" at INL in Idaho Falls.

A passionate supporter of recycling nuclear waste, Dr. Forsythe wrote this book to allay the public's fear of everything nuclear and present the realities of the benefits to the United States of embracing nuclear as the best source of energy to power its electricity generating plants. A co-founder and Vice President for Business Development of Learning Curve Coalition, Inc., an education and management consultation company, she is a professor teaching graduate-level courses in the Florida Institute of Technology's master's degree in engineering management (M.S.E.M.). Dr. Forsythe is a life-long educator.